Naturalising Badiou

Naturalising Badiou

Mathematical Ontology and Structural Realism

Fabio Gironi
IRC Postdoctoral Research Fellow in Philosophy,
University College Dublin, Ireland

palgrave
macmillan

First published 2015 by
PALGRAVE MACMILLAN

Palgrave Macmillan in the UK is an imprint of Macmillan Publishers Limited, registered in England, company number 785998, of Houndmills, Basingstoke, Hampshire RG21 6XS.

Palgrave Macmillan in the US is a division of St Martin's Press LLC, 175 Fifth Avenue, New York, NY 10010.

Palgrave Macmillan is the global academic imprint of the above companies and has companies and representatives throughout the world.

Palgrave® and Macmillan® are registered trademarks in the United States, the United Kingdom, Europe and other countries

ISBN: 978–1–137–46346–3

This book is printed on paper suitable for recycling and made from fully managed and sustained forest sources. Logging, pulping and manufacturing processes are expected to conform to the environmental regulations of the country of origin.

A catalogue record for this book is available from the British Library.

A catalog record for this book is available from the Library of Congress.

To Jasper, my companion at the bonfire

Contents

Acknowledgements

I would like to enumerate a long and impressive list of fellow students and prominent scholars to thank for the priceless insights I have drawn from long conversations with them, over late-night glasses of single malt and endless walks in the misty countryside. However, this work is mostly the product of solitary struggle, mainly involving sparkling water and a comfy armchair, and such a list is thus very short indeed. Not being the most conversational type, my most productive work is done in isolation, with the sole silent, yet not mute, company of books.

Of the few individuals who have been instrumental in the successful completion of this book my gratitude goes, first of all, to Chris Norris. As one of the handful of scholars worldwide capable of following and assessing my conceptual journeys across the lands of analytic and continental philosophy, his guidance has always been attentive but unobtrusive. He has time and again frustrated my hopes to try and refer to a book he hadn't already read, he has voiced his disagreements with some of my ideas as a peer, without ever pressuring me to recant them, and he often seemed to have more faith in my project than I ever did. The breadth of his interests and knowledge, coupled with his humble and unassuming attitude towards everyone, has been an outstanding example of scholarly virtue to me. I have been immensely privileged to be his student.

For good or bad, my intellectual trajectory was radically modified by the discovery, sometime in early 2009, of a highly active internet community of philosophers that was flourishing around the then-novel movement of 'speculative realism'. Whatever the intrinsic merits of this dubiously unitary movement, the virtual encounter with many bright young thinkers loosely associated with it has been tremendously important for the development of my thought. It forced me continuously to try to catch-up with brighter minds, a chase (still in process) which time and again shattered my fragile self-esteem, and punctuated by the occasional, dreaded gaze into the gaping abyss where The Books You Haven't Read Yet lurk. In one way or another several of these dozens of virtual acquaintances have influenced me. Inevitably, however, some of these had a greater impact on my philosophical commitments, a set that luckily comes to largely intersect with those I now have the privilege to call friends. Pete Wolfendale has for me variously taken on the garb of unreachable role-model of philosophical knowledge and responsibility, tireless unraveller of conceptual thickets, earnest networker, friendly critic and, last but not least, fellow lover of horror and science-fiction movies of dubious quality. Dan Sacilotto has become one of my most regular morning (late-night for him) online chats, when he shares with me his humbling and deeply instructive philosophical insights as well as his unrepeatable opinions about me and my work, and hot-bloodedly supports my complaints against more or less fashionable figures or trends in our intellectual environment. Paul Ennis was one of my earliest online acquaintances, standing out among an occasionally snide and pretentious crowd (the dark side of online communities) for his intellectual generosity and candid enthusiasm for novel and daring philosophical ideas, something that took quite concrete shape when he created *Speculations* and accepted me (at a time when we had never even met) as co-editor. Editing the journal with him and the other guys on the editorial team has been an incredibly formative experience, as well as offering me a venue for publication and an occasion to interact with some of the best scholars in our field.

Adrian Johnston has read my work with painstaking attention, and given me flattering praise and insightful critique: I am extremely grateful for his encouragement and support. Lee Braver has provided commentary on a chapter of this book and has flattered me on more than one occasion by asking me for comments on his own work. My thanks goes also to Ray Brassier, who has had a significant, if largely indirect, influence on my recent philosophical interests, and has always been exceptionally kind and supportive in our occasional email exchanges. Naturally, none of these individuals bears any responsibility for (nor do I expect them to agree with!) the contents of this book.

My heartfelt thanks goes to Aaron Swartz, whose proper name will metonymically indicate the militant multitude of anonymous individuals who made my research so incalculably easier.

Finally, my family. My parents and my brother have been and remain enormously supportive of my choice (or should I say my *hope*) of an academic career, have been patiently enduring long periods of silence, and have all performed loving child-caring duties, allowing me a priceless few distraction-free hours of writing. Sîan deserves more credit than I can fully express here, having lovingly and selflessly provided me with indispensable emotional, financial and editorial support. I thank her dearly for her patience on those numerous occasions when my mood was pushed way beyond my standard levels of mild cantankerousness into the (even) less pleasant territories of depressive self-doubt and frustration-induced grumpiness. It's a shame that all the witty ways to say that such a book would just not have been possible without her have been already used and abused. But there you have it.

This book is dedicated to Jasper, or 'Ming' as we affectionately have taken to calling him since his birth, three and a half years ago. A dedication is a pretty poor compensation for all the times I have had to give attention to my laptop or my tablet instead of helping you with your painstakingly careful parking manoeuvres of cars and helicopters on the coffee table, or when I neglected to take sides in the epic battles between diggers and dinosaurs taking place on the carpet. I fear that the sight of me sitting and typing on a keyboard (or the future's equivalent of one) will become all too familiar as you grow up. Perhaps one day you'll help me out with that. While breaks taken with you have helped replenish my intellectual vitality, your contribution to this book, I am afraid to say, has mostly been disruptive: I hope you'll keep it in mind and consider it as mitigation when, in a few years from now, you'll read bits and pieces of it and find that your dad did a rather botched job. I have no doubt that you'll do much better than this. I love you very much.

Note about Citations

All italics in citations are present in the original text unless otherwise stated. I have occasionally corrected 'man' to 'human' in citations, except where this required a too radical syntactic rearrangement (for example, several examples of 'his' or 'him' to replace). Sexism isn't charming, even when retro.

With two exceptions, when citing classical or historical texts (works of Kant or Spinoza, for example), I refer to both the page number of the edition in my Bibliography and to the standard referencing (a book and line/paragraph/proposition reference). The first exception to this rule is Aristotle's works, which I cite by following conventions and referring to Bekker numbers alone. The second exception is Peirce's *Collected Papers*, for which I have given no page number, but only the standard '*CP*' referencing. The *Collected Papers* are available online, and the original eight volumes published by Harvard University Press between 1958 and 1966 are now out of print and extremely pricey.

In citations, I have preserved the capitalisation (or lack thereof) of the noun 'Being' (or 'being') as in the original text: the capitalisation of this word is extremely inconsistent, especially in translations (of Badiou, for example), but I have preferred to leave it unchanged. When used in my own voice, the noun is always capitalised (for somewhat polemical reasons that will emerge through my arguments). The same goes for the italicisation of the Latin '*qua*' and '*a priori*', which I personally prefer italicised, but have left them as they were in citations.

Introduction

This book aims to offer a naturalist correction – something between a creative misreading and an unsolicited deliverance – of Alain Badiou's philosophy, under the aegis of a metaphilosophical commitment to the overcoming (or *neglect*) of the artificial, deleterious and obsolescent division between 'continental' and 'analytic' approaches to philosophy.[1] I will occasionally employ the term 'synoptic' (as in 'synoptic philosophy' or 'the synoptic philosopher') to index this kind of philosophically ecumenical attitude. I am convinced that, in particular in the philosophy of science, we should tirelessly endeavour to relinquish these capricious and myopic restrictions to the gamut of acceptable philosophical assets, and encourage philosophers to develop the ambidexterity[2] to evaluate them competently.

In accordance with these commitments, I shall stage an encounter of Badiou's mathematical ontology and theory of truth with contemporary trends in philosophy of science and philosophy of mathematics: a gap-bridging exercise in line with Badiou's own ambitious (but arguably unfulfilled) claim that his work should be read as marking 'the nullity of the opposition between analytic thought and continental thought' (2006a: xiv). The most general aim of my effort is to target latent theological presuppositions – those principles the genealogy of which is inextricably linked to (that is, in metaphysical connivance with) a theistic worldview – by fusing (and mutually revising) the immanentist and anti-theological potential of Badiou's ontology with a naturalist worldview informed by Anglo-American philosophy.

This exercise is necessarily a speculative one: the often incommensurable vocabularies and concepts require the synoptic philosopher either to take a stance on one side of the divide or to produce her own creative synthesis. I will favour the latter option, motivated by my sympathy towards the project of a *creative metaphysics*. I interpret the latter in accordance with Adrian Moore's[3] definition as 'the most general attempt to make sense of things' (2012: 1) through the creation of new concepts capable of bootstrapping our cognitive frameworks beyond current boundaries. Given that I take philosophy as a whole as an explanatory enterprise striving for the endless enlargement of our understanding of the universe,[4] and our place in it (or, in William James's remarkable formulation as reflecting 'our individual way of just seeing and feeling the total push and pressure of the cosmos' (1955: 18)) I hold that naturalist diligence without conceptual creativity is short-sighted and sterile, whilst conceptual creativity without naturalist diligence is self-referential and frivolous. Most generally appraised, then, this whole book is a syncretic exercise aimed at offering a naturalist metaphysics – a way to make sense of things – which does not rely on any *theological given*, and more specifically a metaphysics capable of making *fully immanent* sense of our scientific enterprise. Epistemic (trust in their method of sense-making) and ontological (confidence in their theoretical content) commitment to the natural sciences does not entail wholesale rejection of metaphysical speculation.

The first chapter offers a broadly sympathetic overview of Badiou's ontology, presenting its main set-theoretical terms, focusing in particular on his treatment of the Void and the Infinite and highlighting the secularising and demystifying momentum of this system.

The second chapter singles out the most troubling shortcoming of Badiou's system, one that threatens to undermine the anti-theological import of his ontology: the inability to offer an intelligible account of the relationship between the ontological and the empirical. In the first, bottom-up attempt to circumvent this problem, I make a foray into the debates in the philosophy of mathematics (against his conviction that his position is able to completely undercut any such debate) and the cognitive neurobiology of the acquisition of mathematical concepts, to demonstrate how our mathematical abilities can be thoroughly demystified. Additionally, I survey contemporary responses to the problem of the applicability of mathematics to the natural sciences. The chapter closes by enumerating possible answers to this problem and introducing the structuralist identification of the mathematical and the physical (my top-down approach) to be fully pursued in Chapter 4.

The third chapter takes a momentary step back from the main argument and engages in detail with the broad problems of scientific realism and naturalism, and the contested relationship between science and metaphysics: herein I present my synoptic commitment (again with reference to 'continental' and 'analytic' sources alike) to both a naturalised metaphysics and a historicist-friendly,[5] post-Kantian attention to the evolution of *a priori* conceptual structures. I further highlight how the primary task of any realism is to offer a science-informed yet responsibly speculative picture of the universe, free from vestigial (onto) theological notions of a divinely ordained order of nature or pre-established harmony between our socio-historically evolving, concept-laden epistemic efforts and the mind-independent reality that they attempt to describe.

The fourth chapter is dedicated to the second, top-down attempt to offer a naturalist supplement to, or correction of Badiou's philosophy, what I will call a 'matherialist' worldview. Opening with a brief introduction of the debate over scientific realism in the philosophy of science I offer a detailed overview of contemporary structural realism (in both its epistemic and ontic forms), and propose a speculative encounter between this stance and the recent structuralist current in the philosophy of mathematics. My ultimate objective is that of a non-reductionist erasure of the distinction between the abstract/mathematical and the concrete/physical, in order to radicalise (and make naturalist sense of) Badiou's mathematical ontology.

The fifth and final chapter weaves together the threads of my argument and shows how a (modified) anti-constructivist Badiouian stance can supplement the structuralist worldview presented in Chapter 4, in accordance with the realist and naturalist commitments developed in Chapter 3. Rejecting Badiou's unfortunate residual Heideggerianism and his concern with a multiple-Being beyond presentation, I defend the matherialist position that there is nothing more (nor less) to reality than structure against accusations of Pythagorean idealism. I will then offer a demystified adaptation of Badiou's notion of truth procedure as an asymptotic rational process of abductive discovery. What is finally preserved of Badiou's approach is the metaontological weight he puts on twentieth-century metamathematical results and on their description of a formally incomplete mathematical reality. I will substantiate this insight by arguing that an ontology where randomness and the transgression of limits are immanent phenomena cannot be completely captured by foundationalist programmes and is intrinsically resistant to any theological re-appropriation.

Having sketched the contents of this book I need now to make four clarifications regarding my approach to Badiou's oeuvre. First, I will not, in this book, offer a complete overview of Badiou's thought, since that would merely (and uncreatively) repeat the vast (and still growing) secondary literature on his work, including some excellent book-length accounts which I could hardly improve upon.[6] My reading of Badiou is of course influenced by the work of previous interpreters and I will critically discuss some of these readings in what follows, but I believe that the time has come for Badiou's work to be creatively developed, reinterpreted and criticised rather than merely commented upon.

Second, it can be argued that, of the triad of core concepts Badiou himself presents as orienting his entire thought – Being, Truth and the Subject – it is the latter that, when the theoretical dust settles, stands as his main concern, largely because it is the *political* dimension/consequences of his system that he most zealously upholds[7] (from his early work onwards Badiou endeavoured to supplement the perceived shortcomings of Althusserian structuralist Marxism with a fully developed, theory of the autonomous political subject of Sartrean/Lacanian inspiration). The Subject, however, is also the concept that figures least prominently in this book. I do not believe Badiou's own account of subject-formation to be particularly persuasive, given his resistant attitude towards naturalist perspectives (and indeed his penchant for quasi-supernaturalist ones), his much-vaunted materialism notwithstanding. There is a causal connection here: his excessive emphasis on the role of an intrinsically non-naturalisable *surgissement* of militant subjectivity is the reason why he forsakes broader naturalist explanatory projects. I have no hostility to attempts to theorise the genesis and nature of rational agents and indeed I remain convinced that any naturalist world-view cannot be complete without a place on the physical continuum for norm-bound subjects; yet I reserve the need to further pursuit this project – that of squaring a natural world of causes with the normativity proper to conceptual rationality, eschewing both the orthodox or 'bald' naturalist dissolution of the latter into the former and the 'liberal naturalist' postulation of a *sui generis,* non-supernatural yet non-scientific domain of reasons[8] – in my future work. While it could be objected that I underestimate how the opening, ontological, meditations of *Being and Event* implicitly depend on the latter ones, focused on the emergence of subjectivity – to the extent that the very decision of equating mathematics and ontology is taken by a subject (Badiou) walking the militant path of a truth-sequence – I want to underline how one of the aims of this book is precisely that of demonstrating that we can preserve (revised formulations of) Badiou's claims on Being and truth without being bound by Badiou's (limiting, in my view) account of subjectivity.

Third, in what will follow I pass over most of Badiou's 'later' work, that developed between the late 1990s and the mid-2000s, and which found a systematic form in his 2006 *Logiques des Mondes* (translated in 2009 as *Logics of Worlds*). The reason is simple. The primary aim of *Logics of Worlds* is to supplement the ontological system of *Being and Event* with a 'phenomenology' or theory of appearances, formalised by means of category theory, capable of accounting for the varied intensities of existence of non-ontological multiples in determinate 'worlds', and to demonstrate a continuity between the minimalist set-theoretical ontology of *Being and Event* with a category theoretic 'logic' or theory of relations between appearing – empirically identifiable – multiples. Given that I offer an alternative, and incompatible, solution to the same problem of translating a mathematical ontology upon the empirical realm, I believe that, once (and if) my naturalising 'correction' of Badiou's ontology is accepted, the complex phenomenological supplementation excogitated in *Logics of Worlds* becomes redundant. In addition, and aside from the merits of my own account, it is eminently debatable whether Badiou's solution is a satisfactory one in its own terms: this is reflected in the somewhat cold reception (certainly as compared with his previous work) his second *magnum opus* received in the secondary literature.[9] As Daniel Sacilotto has very clearly put it, with *Logics of Worlds* Badiou

> has merely transferred the question from the connection between set-theory and the world to the connection between the mathematical duplicity of set-theory and category theory, and the world. If Badiou's mathematical Platonism is not a metaphysical reification of mathematical objectivities, or an idealism claiming for the identity of being and thought, then the connection between the non-mathematical and the mathematical is still pending. (2013: 67)

On these grounds, I will not go into the argumentative details of this latter text, limiting myself to some occasional reference to it where it clarifies the ontological arguments of *Being and Event* or where it offers other interesting observations for my purposes (less systematic and lacking the precise architectonics of its well-structured predecessor, *Logics of Worlds* offers itself readily to this kind of selective approach).

As a fourth and final point, it occurred to me that this book might perhaps have been best titled 'Eviscerating Badiou' since, by its conclusion, I will have rejected virtually every single one of his major tenets. However, what I do preserve, and indeed what have been deep influences on my intellectual formation, are his metaphilosophical insights into the syncretic role of philosophy, the vouching for a necessary return of a robust idea of truth as the breaking out of stale knowledge, the insistence on mathematics' intrinsic secularising power and, perhaps, a certain irreverence towards the contemporary philosophical consensus. What needs stressing, however, is that this is *not* a work of Badiouian exegesis: anyone with a minimal familiarity with Badiou's thought will immediately notice that the very idea of 'naturalising Badiou' is profoundly non-Badiouian. I am well aware of that: my heterodox approach to Badiou is not engendered by a misreading of his ideas, but by a fundamental *disagreement* with many of them. Doubtlessly, some among the most orthodox Badiouians will find my downplaying of Badiou's later works a reprehensible interpretative strategy, and will disagree with my amendment of many of his early tenets. I don't consider this work threatened by such accusations: mine is not aimed to be an interpretation, but a selective appropriation and a revision. As I have already explained, my most general motivation is that of a synoptic and creative encounter of two philosophical traditions: I am not bound by any deference to Badiouian doctrine. One of the main vices of continental philosophy is its penchant for endless, bromidic and intellectually incestuous exegesis of 'master figures'. Very much in Badiouian spirit, I prefer risking the composition of a (potentially wrongheaded) new philosophical stance over the pious and textually sanctioned exposition of established ideas.

Before bringing this Introduction to an end, I want to make a few observations on the background commitments guiding this work, buttressing my choice of terms. First, I have already repeatedly referred to *naturalism*. Over and above the minimal sense of 'rejection of supernatural causes',[10] in ontological terms naturalism is the view that all there is is what the natural sciences describe. In epistemological terms, it is the view that the natural sciences are our best (most reliable) epistemic practices. In semantic terms, naturalism is the view that all vocabularies can in principle be reduced (no matter how long the chain of reductions actually is) to the privileged vocabulary of the natural sciences, particularly physics. Widespread consensus can hide supine acceptance of dogmas: is naturalism such an unexamined assumption? As I have adumbrated above, I will examine the notion of naturalism in greater depth in Chapter 3; for now, I want to answer the latter question in the negative by highlighting how I take *naturalism* to be a commitment to ontological and explanatory *immanence,* and view both as unavoidable consequences of a post-Enlightenment worldview.

This is a tall order. Explaining a contested term by reference to a *much more* contested one seems like an unwise strategy.[11] Minimally, however, my commitment to Enlightenment values, moulded upon the Kantian 'normative turn' most recently and vigorously extolled by

Robert Brandom, implies a fundamental assumption of *responsibility* (against ideological and reductive equations of the Enlightenment with vulgar forms of mastery, instrumental rationality and anthropo-/ethno-/Euro-/andro-centric forms of ideological supremacy).[12] The 'emergence from self-imposed nonage' (to quote Kant's well-known definition of the Enlightenment)[13] should not be collapsed into the self-interested pursuit of mastery over reality or over other human beings, but rather interpreted as the assumption of responsibility for one's claims and actions.[14]

The intellectual maturity and autonomy[15] vouched for by Kant do not trivially amount to a pernicious self-sufficiency of reason as that decried by critical theorists (from Weber to Habermas through Adorno and Horkheimer) and lesser postmodernists[16] – that would merely diagnose a move from infantile immaturity to self-important teenage rebellion. The systematic scepticism and subversion of entrenched beliefs professed by the 'hermeneutics of suspicion' is but a (no doubt necessary) stage towards the higher goal of intellectual maturity: debunking undermines childish naiveté, but as a cynical, relentlessly critical stance it falls short of a fully adult, constructive assumption of responsibility.[17]

Far from arrogant mastery and intellectual invulnerability, the Enlightenment ideal of responsibility (especially as it mandates the explanatory enterprise of science) denotes accountability and indeed a certain vulnerability to failure: *Aufklärung* is more of a structurally interminable process of revision than an irrevocable achievement.[18] The purely *formal* norms of rational engagement with the universe and with each other (the norms guiding the use of concepts in judgements) do not *invest* the agent with authority, but *enjoin* it to act in conformity with them; a conformity that needs to be constantly re-negotiated and can never be taken for granted. Reason's 'interests' are not homogeneous with the interests of the self (either *qua* utility-maximising subject or *qua* survival-seeking biological organism) but are congruous with those of the totality of rational agents, collectively engaged in explanatory projects.

This Enlightenment ideal of responsibility directly impinges upon the arguments of this book (and its anti-theological ambition),[19] by grounding the notion of *autonomous explanation*. If Enlightenment means ceaseless caution against intellectual complacency, our explanatory practices should systematically avoid theological 'explanations' via unaccountable transcendent causes responsible for the apparent order of things in nature.[20] Given that the best of these practices our species has devised are the natural and social sciences, there is something *intrinsically emancipatory* in the very notion of scientific explanation, insofar as it is our intellectual maturity on the line in the fallible enterprise of offering revisable explanations: while science is the paradigmatic case (and where momentous social consequences are at stake, the proper functioning of science is vital), the ideal of an explanatory responsibility can be exported to ethico-political fields, to explicate protocols of accountability for any process of decision- and policy-making.

The epistemic responsibility of an ever-refined understanding of the universe cannot be delegated to anyone or anything but *ourselves* as a community of investigators, and it should ruthlessly push us to renounce any transcendent support. Hence naturalism's explanatory closure: explanations (and scientific explanations as paradigm instances) should be autonomous *qua* committed to immanence – no transcendent causes in the universe, no divine aid to our knowledge. This corrects our understanding of the somewhat worn-out ideal of *disenchantment:* to disenchant the universe does not resolve in a bovine tuning down of our receptivity to the strangeness and wonders it offers. On the contrary, it amounts to the determined attempt to take upon ourselves the responsibility to offer revisable (and perhaps never *fully* completed) and intersubjectively appraisable explanations for them, resisting the temptation to throw our hands in the air calling out for a miracle. Scientific thought is the practice of disentangling seemingly intractable – and at first sight mysterious – phenomena until an explanatory strategy is identified, through the implicitly social exercise of reason, to elucidate them. It follows that I will employ the term *secular* as essentially synonymous with 'committed to immanence' and with being 'constitutionally open to public scrutiny', so that 'secular' explanatory practices or metaphysical assumptions are those in line with an immanentist worldview and products of the employment of our collective rationality.[21]

1
Badiou's Mathematical Ontology

In this chapter, I offer a reading of the central nodes of Alain Badiou's mathematical ontology of multiplicity, highlighting those elements which I take to be most significant for the secularising naturalist, and paving the way for my attempt to link Badiou's philosophy – his mathematical ontology (here) and his 'generic' conception of truth (Chapter 5) – with contemporary trends in analytic philosophy of science. It is my contention that Badiou's metaphilosophical insights, properly reinterpreted and recast in the service of naturalism, can be of use for a fully secularised metaphysical worldview. In the course of this reading I will encounter an obstacle offered by Badiou's philosophy, that of the unsettled relationship between the ontological and the ontic, empirical realms.

The strongest and philosophically most fertile claims of Badiou's philosophy are the two main consequences of his equation of mathematics with ontology: that 'the one *is not*' (Badiou 2006a: 23) (a thesis about Being, or what there is) and that the infinite can be secularised (or de-romanticised, made thoroughly immanent) – and has been, thanks to the mathematical work of Georg Cantor – by rationally demonstrating its thinkability (a thesis about truth).[1] These two theses – two *axiomatic decisions* at the base of the set theoretical universe which Badiou adopts as an ontological model – while undermining both the infinitely other God of the prophets (or phenomenologists) and the self-present God of the metaphysicians, can deliver to the scientific realist two authentically post-theological metaphysical principles.

1.1 Death(s) of God: for axiomatic immanence

The goal of any anti-theological philosophy is immanence, that is, the equivalence of the ultimate ground of Being with the totality of its empirical instantiations. Indeed, the desired outcome of *any* contemporary philosophy (more so when committed to naturalist guidelines) must be the articulation of an immanentist metaphysics: I take the repudiation of a transcendental ground of sense as being not merely an historically contingent philosophical turn, or even timely, but rather an unavoidable feature (and an unrelinquishable achievement) of human intellectual development. As Deleuze and Guattari eloquently put it, '[i]mmanence can be said to be the burning issue of all philosophy because it takes on all the dangers that philosophy must confront, all the condemnations, persecutions, and repudiations that it undergoes' (1994: 45). Atheism, now a fashionable label for any self-appointed defender of rational thought, must be philosophically articulated as conceptual immanentism.

Alain Badiou's most explicit atheist pronouncements are to be found in the prologue of his *Briefings on Existence* (Badiou 2006c). Here, interrogating the equivocity of the referent of the Nietzschean statement 'God is dead!', Badiou draws a distinction between three different understandings of God. According to him, the history of metaphysics itself is the history of 'the mortification of God' (2006c: 25) since the *metaphysical* concept of God running from Aristotle to Descartes (and indeed Kant) is, precisely, a *concept* in light of which, the philosopher can make sense of things.[2] Such is the God of the philosopher, frozen, by Galileo and Descartes 'in the trans-mathematical punctuality of actual infinity' (2006c: 22), working as a Principle allowing for the 'suture of mathematical truths to their being, or [as] a guarantee of judgments in the shape of the Other' (2006c: 23). The infinite is linked to the One in the concept of a unique God of metaphysics.[3]

Such a God, Badiou argues, is a God who never lived, a far cry from the God of religion, the God of Paul, of Pascal and Kierkegaard, a 'living God which is always *somebody's* God' (2006c: 23) met through personal, lived encounter and not through logical proof.[4] As Pascal had it: '[t]he God of Christians does not consist in a God who is merely the author of geometrical truths and of the order of the elements' (2004: 227). In aphoristic form, Badiou summarises this opposition claiming that '[w]ith respect to God, it is true that religion is vivifying and metaphysics mortifying' (2006c: 25). Therefore, even though the death of God means that 'religion is finished' (2006c: 23), this death 'leaves the question of the destiny of the God of metaphysics unresolved' (2006c: 26). The evacuation of the latter from philosophy began with Nietzsche (whose announcement of the death of God was, in Badiou's taxonomy, the paradoxical announcement of the death of a dead God, the God of

Metaphysics), but was only accomplished by Heidegger, having identified Nietzsche's reversal of Platonism – becoming over Being – as the swansong of Western metaphysics. His *destruktion* of onto-theology amounted then to the rejection of a univocal understanding of Being (and God) as a metaphysical concept of the highest of beings. Yet, for Badiou it is Heidegger himself who opens the space for the third kind of God (what Badiou calls 'Heidegger's aporia' (2006c: 27)), the God of the Poets. This is the God with which the philosopher-poet stands in a 'nostalgic relationship' (2006c: 28) of epochal expectation. Through a 'meta-poetic metaphorizing' (1999: 43) the post-metaphysical philosopher hopes for a re-injection of meaning into the world and orients his own finite being towards an attentive but passive receptivity to the historical self-presencing of Being.[5]

In sum, while the living, personal *God of Religion* has been repeatedly killed off – in a dialectical evolution running throughout the history of Western philosophy – by the aloof *God of Metaphysics*, it remains for us to free our ontology from the latter's constraints while being wary of the nostalgic invocation, by those pronouncing the end of metaphysics (and its God) under the banner of existential finitude, of a *God of the Poets*. This latter God – currently the most insidious one – is expected as the only hope of a re-enchantment of the world by those finite, death-ridden mortals incapable of autonomous production of meaning and of mastery over the world (nihilistic mirages, as Heidegger considered both Nietzsche's philosophy of the Will and humankind's techno-scientific ambitions). Truth has vanished with the God of Metaphysics, and the philosopher can only have faith in the (non-philosophical) promise of a meaning to come.[6] It is at this historical juncture that it is necessary to reaffirm a contemporary form of atheism as the destitution of all three varieties of God, one that is 'all in all ... about finishing up with promises' (2006c: 29) and which decisively breaks with a philosophy 'caught between the depletion of its historical possibility' and 'the conceptless advent of a salvational turnabout' (1999: 114) that is, the poetic[7] hope for a redeeming return of (a) God.

Such an atheism pivots upon the decoupling of meaning from truth,[8] and of the one from the infinite, in order to 'restitute the infinite to the banality of manifold-being' (2006c: 30). What needs restating is the

> great modern declaration: the infinite exists, and, what is more, it exists in a wholly banal sense, being neither revealed (religion), nor

> proved (mediaeval metaphysics), but being simply decided, under the injunction of being. (2008a: 82)

Elsewhere, Badiou states that it is 'necessary to de-suture philosophy from its poetic condition. Which means that it is no longer required today that disobjectivation and disorientation be stated in the poetic metaphor. Disorientation can be *conceptualized*' (1999: 74). His atheism, in other words, amounts to the proclamation of a philosophy of *axiomatic immanence*[9] capable of mathematically articulating the infinity of Being without attributing to it *any* meaning-full, transcendent significance.[10] The immanentist creed is thus encapsulated in the aphorism (both declarative and imperative) that 'everything is here, always here' (2006c: 31). The atheist philosopher rejects the metaphysical *pathos* of (a) reason without (poetically) forfeiting the power of Reason. Immanence signifies responsibility, rational accountability *qua* the abandonment of any heteronomous intervention in (or justification for) either our rational life or natural reality, and it informs the anti-dualistic acknowledgement that the former is part of the latter.

This immanentist, or simply *philosophical*[11] stance,

> is in rupture with [Nietzschean] anti-philosophy because it both retains and develops, by means of a rational critique, the idea of truth. But it is also in rupture with religion, because it refuses to identify truth with sense; it even willingly declares that in any truth there is always something of the nonsensical. (2001c: 9)

To champion atheism means to defend an *immanentist ontology* and a *secularised epistemology*. And this, as we will see, is possible only for a philosophy developing between the twin poles of the axiomatically established existence of the void and the infinite.

1.2 Subtractive ontology

Axiomatic decisions, for Badiou, are the anti-historicist duty of contemporary philosophy.[12] His rejection of the poetic approach to philosophy and its nostalgia for the divine, accordingly, begins with a founding axiom: the equation of mathematics with ontology.

Herein lies the most delicate point of Badiou's project: the equation mathematics = ontology is a *decision*, an unprovable, foundational axiom,[13] which can be evaluated only by its *consequences*. Badiou adopts an axiomatic procedure in explicit opposition to the poetico-linguistic philosophical climate, perceived by him as dominating the contemporary (in late 1980s France) intellectual scene.[14] Unlike poetry, for Badiou mathematics

> disciplines thought through explicit rules, not through the singular genius of language, and offers to everyone a shared demonstration, whilst never giving up on ultimate clarity – as complicated as its construction may be. It informs the True without conceding anything to the trembling or existential doubt before that whose cruel necessity it unveils. (2006d: 40)

However, this methodological point is hardly enough to warrant the equation of mathematics with *ontology*, an equation expropriating philosophy of one of its historically most central branches and recasting the philosophical role as the non-self-sufficient compossibilisation of non-philosophical truths. Could the unsympathetic reader question the validity of this axiom and reject it, thus collapsing Badiou's system upon itself? The answer, quite simply, is *yes*.[15] Axioms, in mathematics as elsewhere, are a matter of pure choice, posing unquestionable logico-architectonic constraints on all that which inferentially follows, yet incapable of self-justification. Such is Badiou's rationalist wager: the singling out of an event (the Greek 'discovery' of the matheme),[16] and the identification of this event as the starting point of a truth procedure (pursuing the truth of the equation of mathematics and ontology) which will achieve full maturity only with Georg Cantor's creation of set-theory and whose meta-ontological importance will be assessed in his own work.[17] In this light, Badiou's project from *Being and Event* onwards can be read as one long retroactive explanation of the validity of his original axiomatic choice. To this effect, Brassier and Toscano's definition of Badiou's thought as an 'aleatory rationalism' (2006b: 260) is appropriate: his method is that of an inaugural, ungrounded, contingent philosophical choice (a pure choice that is, before anything else, a rejection of traditional ways of mooring philosophy to either an empirico-phenomenal world or to a transcendental subject), whose consequences have nonetheless to be pursued according to the strictest standards of logico-mathematical deductive reasoning. Only an empirical grounding would seem capable of vindicating Badiou's inaugural decision, but at the price of a reinterpretation of Badiou's own views. Herein lies the hermeneutic challenge of this book: how, if at all, can Badiou's radically rationalist (and indeed anti-naturalist) stance be reconciled with a scientific realism which hopes to be warranted by the empirical success of the natural sciences?

The inaugural axiom of Badiou's philosophy, taking a stance on the traditional metaphysical quandary of the priority of the One or the Many, is that 'the one *is not*' (2006a: 23). This should be immediately coupled with its corresponding positive claim: 'what *is* is multiple', or better yet 'Being is multiplicity'. The phenomenal appearance of unity is illusory. What there is, what is *presented*, are structured situations, collections of multiplicities organised by an impersonal operation[18] of *count-as-one* which makes heterogeneous multiplicities consist in a composed unity. In its foundational denial of ontological one-ness such an ontology is intrinsically anti-theological ('there is no God. Which also means: the One is not' (2001a: 25)), and is on a continuum with the denunciation of onto-theology, insofar as Badiou's confinement of oneness as the mere product of an operation of counting-as-one of a multiplicity is in accordance with the deconstruction of an eternal plenitude or self-presence of Being. The One, for Badiou, is always an *effect*, 'whose fictive being is maintained solely by the structural retroaction in which it is considered' (2006a: 90) – not unlike presence, for Derrida, can always be demonstrated to be dependent on an effaced trace.

This dual commitment signals Badiou's deeply original aim of reconstructing a rationalist ontology under the aegis of Plato and Descartes without, however, forsaking the tradition of the post-Nietzschean critique of metaphysics:

> [w]e cannot revoke the without-being of the One nor the limitless authority of the multiple. God is truly dead, as are all the categories that used to depend on it in the order of the thinking of being. The pass that is ours is a *Platonism of the multiple*. (1999: 103)

The operation of the count-as-one, or *structuring*, itself establishes a split between that which logically precedes it as 'having been' gathered together, an *inconsistent multiplicity* – that Being which can only be retroactively discerned in the temporal form of the future anterior – and the *consistent multiplicity* which the situation (any unit-effect composed of multiples) now instantiates in what is presented. Inconsistent multiplicity, then, cannot simply be interpreted as 'the Many', since such a domesticated multiplicity (an indefinite iteration of 'ones') pertains only to that which is presented (as One-ness): Being is neither one nor multiple, but an uncountable pure multiplicity beyond reckoning (thus inconsistent) which, insofar as it 'is', is always subtracted from presentation.

This crucial distinction is explicated by Badiou in an audacious and penetrating reading of Plato's *Parmenides*. According to Badiou, in this text (unique in the Platonic *corpus* for the way in which a young Socrates is caught in an aporetic contradiction by the old Parmenides) Plato found himself pushed to the logical limits of his own doctrine and forced to think inconsistent multiplicity when pursuing the inevitable ontological consequences of the non-existence of the One. In the *Parmenides*, Plato

> formulates an essential ontological truth; that in absence of any being of the one, the multiple in-consists in the presentation of a multiple of multiples without any foundational stopping point. Dissemination without limits is the presentative law itself... [t]he essence of the multiple is to multiply itself in an immanent manner. (2006a: 33)

Crucially, however, this multiplicity of multiplicities cannot be thought as 'the Many': through a meticulous textual deconstruction Badiou locates a point of tension in Plato's terminological distinction between the terms πλῆθος and πολλά. Badiou – implicitly laying claim to this most authentic Platonic heritage – draws a parallel with his own ontology explaining that

> the term πλῆθος; should be thought as designating the inconsistent multiple, the multiple-without-one, pure presentation, whilst πολλά designates the consistent multiple, the composition of ones. The first is subtractive with regard to the one; not only is it compatible with the non-being of the one, but it is only accessible, be it within a dream, on the basis of the ontological abrogation of the one. The second term, πολλά, supposes that a count is possible, and thus that a count-as-one structures the presentation. (2006a: 35)

For Plato, however, it was impossible to think systematically about the wild dissemination of the hypo-immanent inconsistent multiple unleashed by the ontological abrogation of the one. Hence Badiou's reference to Plato's fugacious insight into inconsistent multiplicity as a 'dream' (2006a: 36): Plato lacked the conceptual arsenal necessary to *think* the presentation of multiples from the bottomless inconsistency of Being, that battery of concepts that will be delivered to humankind only with the development of adequate mathematics – set theory – because it is only 'with regard to sets [that] there can be neither a universal set, nor All, nor One' (2000a: 3). Inconsistent multiplicity is therefore that multiplicity subtracted from any individuation that can only be named *after* it is made consistent via its presentation in situations: it thus cannot function as either as a transcendent principle or as a virtual plane of immanence. It is this conception of presented Being as an operational coalescing of a void which allows Badiou to create an ontology open to radical yet immanent novelty, for the void of every situation will be (as I will explain in Chapter 5) where the possibility of an event lies.

So why mathematics? Because mathematics can offer an asubjective *inscription* of Being, not Being itself; it can refer to Being without objectivising it:

> [t]he thesis that I support does not in any way declare that being is mathematical, which is to say composed of mathematical objectivities. It is not a thesis about the world but about discourse. It affirms that mathematics, throughout the entirety of its historical becoming, pronounces what is expressible of being qua being. (2006a: 8)

Put otherwise, the claim that mathematics is the axiomatic presentation of an inconsistent multiplicity – what can be thought of Being – does not imply that Being itself is mathematical. But it *does* imply that there is no opposition between a realm of mathematical entities 'in-themselves' and a thought which encounters them: mathematics is the purest articulation of the thinking of Being itself. Mathematics = Ontology, then, is neither a thesis about Being nor is it a thesis about a Platonic realm of transcendent entities, or of a Husserlian field of transcendental ideal objects. It is a 'thesis about discourse', that is, a mathematical discourse (specifically, set theory) capable of saying the sayable (or thinkable) about Being *qua* Being as presented through content-independent formalisation. Ontology/mathematics, so to speak, sets the rules for the inconsistent multiplicity of Being to be presented/counted as one.[19] Ontology is 'the thought of the inconsistent manifold, that is, what is reduced without an immanent unification to the sole predicate of its multiplicity' (2006c: 40). Mathematics is the formalised thought capable of axiomatically presenting/unifying an intrinsically unpresentable (because inconsistent) multiple Being.[20] Only an axiom 'prescribes without naming' (2006c: 38), that is to say, can lay down structural conditions *irrespective of any content* which these conditions organise, or indeed in the absence of any such content.[21] Badiou explicates the non-mathematicity of Being itself via a distinction between (capitalised)

Number and (lower-case) numbers: while the latter term indexes the operational, structural understanding of numbers (including the rationals, integers, reals and so on) the former is a 'macro-field' of pure inconsistency. Number is 'the inconsistent site of Being for the consistency of numbers', while numbers 'carve out consistent numerical situations within this inconsistency' (2006b: 63). Ultimately then, 'the numbers that we manipulate are only a tiny deduction from the infinite profusion of Being in Numbers' (2008a: 211), a 'minute part' of what Number *qua* multiple-Being is *'capable'* (2008a: 112). The infinite, disseminated field of Number (the inconsistent multiplicity of Being) is not intrinsically mathematical, but can only be thought (made consistent) in mathematical terms. Mathematics is 'the capture of the void and the infinite by the letter' (2005c: 143).

Armed with this understanding of mathematics, Badiou rejects any duality of mathematical subject (the thinker) and object (the thought) by identifying the latter in the immanent 'material' of the former's structure of thought. For Badiou, then, 'we submit ourselves, as every philosophy must, to the axiom of Parmenides: it is the same to think and to be' (2009: 99). Gillespie insightfully notes the parallel between Badiou and Spinoza, as both thinkers aim to construct a rationalist ontology along axiomatic lines: '[f]or each thinker, thought's ability to posit axioms of being is tantamount to an equivalence between thought and being' (2008: 26).[22] As Fraser synthetically puts it, the rationalist equation of ontology and set theory means that '[t]o think a particular structure "in its being" – that is, according to its pure presentational form – is simply to think it in the symbolic language of set theory' (in Badiou 2007a: xlii). Badiou's aphorism, then, should not surprise: 'mathematics does not understand the meaning of the claim "I cannot know"' (2006b: 17) since to know truly is to know mathematically, and all that *is* is expressible in purely symbolic mathematical inscriptions.[23] Clemens explains that '[m]athematics *is* the place of inscription of Being; the letter of mathematics *are* directly ontological' (2005: 100) and Brassier argues that 'Badiou's account is one in which there is nothing that science cannot know because, dispensing with every vestige of substance, Badiou's formalist ontology leaves no room for the inconceivable or unconceptualizable' (in Bartlett and Clemens 2010: 72).

With this de-presencing insight into the ontological relevance of mathematics in place, Badiou terms the 'Great Temptation' that kind of thought which assumes that 'only an experience situated beyond all structure will afford us an access to the veiling of being's presence' (2006a: 26), that is, the temptation of promoting an ontological discourse capable of cutting through presentation and dipping into Being itself. This takes the form of negative theologies, of philosophies of *conatus* towards a One-beyond-Being (Neoplatonism), of subjection before the absolute Other beyond the face (Levinas), or those attributing to language some quasi-mystical power of poetic evocation of Being (Heidegger). Badiou has no truck with these transcendent openings, all of them ultimately signalling a nostalgia for the One: he is adamant in his insistence that 'no access to being is offered to us except presentations' (2006a: 27), that what is presented is always a one-ified multiplicity (not a One) and that ontology is precisely the presentation (in mathematical inscriptions) of these presentations (yet still a *presentation*, not an ecstatic silence). Nothing can be discerned of inconsistent multiplicity in its (pre-count) self, and Badiou is very clear when he explains that

> what is required is that the operational structure of ontology discern the multiple without having to make a one out of it, and therefore without possessing a definition of the multiple. The count-as-one must stipulate that everything it legislates on is multiplicity of multiplicities, and it must prohibit anything 'other' than the pure multiple – whether it be the multiple of this or that, or the multiple of ones, or the form of the one itself – from occurring within the presentation that it structures. (2006a: 29)

Badiou's project is to be located within the philosophical tradition of fundamental ontology. His concern, the question which provokes his thought, is well rehearsed: how can we talk about *Being-qua-Being*? His innovative answer (contra Heidegger, as the last philosopher who asked the question programmatically)[24] is that we have never stopped doing it, but that if we were to look for ontology in the philosophers' work we would be mistaken: it is mathematics, at least since the time of Plato (even though Plato lacked a refined enough mathematics), that has been the discourse of Being-*qua*-Being. Subscribing to Heidegger's preoccupation with the ontological difference (between Being and beings), for Badiou mathematics, 'inscribes being as such' (2006b: 12) as opposed to any other intellectual discipline whose objects are determinate, *ontic* beings. This follows from the warning against any definition of the pure multiple: ontology can only begin *in medias res*, starting from within the situation. And what can *inscribe* Being/inconsistent multiplicity without *describing* it is only an axiomatic system, which Badiou identifies in set theory, that branch of mathematics that studies abstract sets, considered as foundational to the entire mathematical edifice. The adoption of a mathematical ontology of the multiple produces attractive results for those seeking a non onto-theological metaphysics capable of offering a secular grounding for a realist interpretation of science since, as Badiou explains, 'mathematics dismantles the perilous theological connection Truth-Being-One' (1994: 86). Mathematics, in his reckoning, offers a purely formal and aloof *me-ontology* uniquely able, via its axiomatic architectonics, to retroactively present Being *as* structured while denying its one-ness.[25]

Once again, ontology/mathematics is the generic form of pure presentation of an inconsistent multiplicity: insofar as we can discursively articulate Being, we can do so only mathematically. This rationalist confidence of the power of mathematics makes Badiou's claim of *materialism* for his own philosophy seem problematic, but it can be accounted for along two axes: his insistence on the materiality of mathematical inscription and his adherence to 'Lucretian' materialism. As for the former, Ray Brassier explains that

> [t]he 'materiality' of mathematical practice is not to be understood as an analogue of the inexistent philosophical category 'matter,' but rather as an index of the scriptural production of difference. This account of scriptural materiality is, so to speak, the esoteric subtext of Badiou's materialist epistemology of science. (2005: 145)

Thus, Badiou's Platonism must be interpreted as a stance that 'recuses the empiricist distinction between thought and object' (2005: 136), a conviction traceable back to his early, Althusserian (and, by intellectual lineage, Bachelardian)[26] structural-materialist project of disengaging science from its ideological terrain, developed in *The Concept of Model* and his early papers published in the *Cahiers pour l'Analyse*. Ontology occurs in its scriptural (material) presentation, where ontological *thought* and *inscription* are united (through an immanent fusion of the ideal and the material), without referring – as the bourgeois epistemological distinction between the concrete and the abstract requires – to a transcendent realm of mathematical idealities (as, for example, in traditional mathematical Platonism). Mathematics then, is no mere formal tool – extrinsically applied to an empirical substratum – but the *productive* engine of science, operating by differentiating its own scriptural material.

Relatedly, Badiou's admiration for 'the magnificent figure of Lucretius' (2006c: 35) must therefore be understood on the basis of the latter's revolutionary, anti-Parmenidean (and disenchanting) rejection of the oneness of Being in favour of an endless multiplicity of atoms swarming in the void. Badiou thus reads Lucretius through his own materialism of inscription when he explains that 'for Lucretius, all truth established itself from a combination of marks, from a rain of letters, atoms, in the pure unpresentable that is the void' (2004: 105). We should not, however, mistake Badiou's enthusiasm for Lucretius as an endorsement of the Epicurean reduction of all phenomena to atomic 'bits' of matter (for, as such, Lucretius multiplicity remains a multiplicity of 'ones').[27] This means that, as I will explain below, the void set which is the building block (the 'material') of set-theoretical ontology is not Being itself, but indeed is the *name* (the inscription) of the originary no-thingness of inconsistent multiplicity attributed from within the ontological situation (the sphere of production of scriptural discourse). If for Greek atomism (of Leucippus and Democritus, Epicurus and Lucretius) the axiom was 'the universe is nothing but atoms (Being) and void (non-Being)', for Badiou the axiom is 'there is (Being is) nothing but multiples of multiples, that can only be presented through a mathematical thought which begins by assigning a proper ("atomic") name to this inconsistent multiplicity'.[28]

Badiou's adherence to Lucretian (Epicurean) materialism, then, is best understood as motivated by an a-theological, anti-foundational and anti-idealist outlook, privileging scriptural, machinic repetition of 'non-objectal' atoms/mathemes to a 'continuous' materiality infused with the *pneuma* of meaning.[29] This is the aleatory materialism (or materialism of the encounter) of Badiou's master Althusser, an

> *almost completely unknown materialist tradition* ... opposed, as a wholly different mode of thought, to the various materialisms on record, including that widely ascribed to Marx, Engels and Lenin, which, like every other materialism in the rationalist tradition, is a materialism of necessity and teleology, that is to say, a disguised form of idealism. (2006: 167–8)

This has its origins in Epicurus's atomism. The doctrinal differences between the Epicurean's atoms chaotically swirling in the void and Badiou's inconsistent multiplicity as a materialism of groundlessness can indeed be de-emphasised once the common interest in the *contingent origin* (via Lucretius's unpredictable *clinamen*[30] or Badiou's arbitrary count-as-one) of all that is presented is recognised. Slavoj Žižek puts this opposition between theologised idealism and 'secularised' materialism very explicitly when he explains that the difference between idealism and materialism is that for the idealist, we experience our situation as 'open' insofar as we are engaged in it, while the same situation appears 'closed' from the standpoint of finality, that is, from the eternal point of view of the omnipotent and all-knowing God who alone can perceive the world as a closed totality; for the materialist, the 'openness' goes all the way down, that is, necessity is not the underlying universal law that secretly regulates the chaotic interplay of appearances – it is the 'All' itself which is non-All, inconsistent, marked by an irreducible contingency. (2006: 79)

1.3 Set theory

A brief excursion into set theory is now necessary, in order to gather the mathematical concepts employed in Badiou's ontology and understand why Badiou would describe it as 'the greatest effort of thought ever accomplished... by humanity' (2006a: 499), uniquely capable of offering a theory of the pure multiple.[31] Set theory studies sets, or collections of abstract objects. The birth of set theory is due to the revolutionary work of the German (Russian-born) mathematician Georg Cantor, in the latter decades of the nineteenth century. Cantor's achievement was to ground the notion of number in that of set (*Menge* (aggregate) is the German term which Cantor originally employed) and offer, for the first time in the history of thought, a mathematically regulated method to deal with the concept of infinity (reconceived in set-theoretical terms), an idea which had long offered a number of apparently insoluble paradoxes.[32]

In the mathematical work developed through a number of papers published between 1874 and the end of the century, particularly in the 1883 monograph *Grundlagen einer allgemeinen Mannigfaltigkeitslehre* ('Foundations of a General Set Theory'), Cantor demonstrated the existence of a *hierarchy* of infinities, proving that the infinity of the set of natural numbers is less – of a smaller *cardinality* (*Mächtigkeit*) – than the infinity of the real numbers, the latter being an *uncountable* infinity (Cantor in Ewald 1996, Chapter 19). A *countable* infinity, on the other hand, can be put in a bijective function with the set of natural numbers: in other words, an infinity whose members can be put in a one-to-one correspondence with the sequence of natural numbers. Cantor proved, with his celebrated method of diagonalisation,[33] that the set of real numbers cannot be countable (or denumerable), and that therefore the cardinality of the (infinite) set of reals must be higher (uncountable) than the cardinality of the (infinite) set of natural numbers. This profoundly counter-intuitive result opened the doors to a deluge of larger and larger infinities in Cantor's theory of transfinite numbers.

Briefly, these can be divided between transfinite ordinals and transfinite cardinals. With transfinite *ordinals* Cantor expressed the possibility of finding a successor for any infinite number: the sequence of natural numbers can be written as a well-ordered set $\{1,2,3 \ldots \omega\}$ where ω would be the first infinite ordinal[34] number after all the finite whole numbers, but to the ordinal ω there can always be found a successor $\omega + 1$, $\omega + 2$, $\omega + 3 \ldots$ and so on to infinity. These are the transfinite ordinals. But it is clear that whatever transfinite ordinal we consider, its cardinality will always be the same as the set of positive integers; it will always be a countable infinity. So how can we express the uncountable infinity of the reals? Cantor, in later work, introduced the transfinite *cardinals*: with the Hebrew letter $\aleph_0$ which indicated the cardinality of a countable infinity (the cardinality of ω and of all of its infinite successors) and established this as the smallest kind of cardinality, others being arrived at by the power set operation upon the previous one. The formation of the power set P(x) of a given set x produces a grouping of all of the subsets of x in all possible combinations thus creating a new set P(x) to which all of these subsets belong. In finite cases, it is easy to see that this operation obeys the rule that for any set x of cardinality n the power set will have a cardinality equal to 2^n: for a set of three members its power set will have eight members. Generally, the power set of any set will always have larger cardinality than the initial set. Things get more interesting with infinite sets. Cantor proved, via his diagonal argument, that the application of the power set operation to the infinite set ω of all natural numbers will produce another infinite set with *larger* cardinality. If the cardinality of ω (and of all of its transfinite ordinal successors) was identified with $\aleph_0$, the power set operation would produce a larger ordinal to be called $\aleph_1$, and so on. This means that the universe of sets is a hierarchical progression of finite and transfinite ordinals (obeying a simple successor rule n + 1) broken by immeasurable gaps where, through the power set operation, a leap to a larger cardinality takes place.

In his hierarchical picture of the mathematical universe, a problem arises: *where* does this transfinite progression take place? Is there a set of all sets, finite and transfinite alike, which, so to speak, could 'contain' the entire mathematical universe? This notion, as Cantor immediately noted, leads to paradoxes. And the meaning of these paradoxes was clear to him. What he called 'inconsistent multiplicity' the unmeasurable (a multiplicity wherein the assumption of consistency leads to contradiction), absolute(-ly infinite) ground of the transfinite realm was, as far as the pious Cantor was concerned, *God itself*. What lies above the progression of the transfinite is an (actual) infinite unity which encompasses everything but cannot be mathematically grasped, cannot but be God himself. Cantor secularised the infinite only to the extent that he created an 'intermediate', albeit astonishingly large, realm of transfinite numbers. The truly infinite, God, remains a topic for theology, not mathematics. Badiou notes how Cantor's own understanding of set theory was still dependent on a metaphysics of presence and totality, what he calls an onto-theological orientation of thought. Thus, he writes:

> Cantor's ontological thesis is…that inconsistency, mathematical impasse of the one-of-the-multiple, orientates thought towards the Infinite as supreme-being, or absolute…Cantor, essentially a theologian, therein ties the absoluteness of being not to the (consistent) presentation of the multiple, but to the transcendence through which a divine infinity in-consists, as one, gathering together and numbering any multiple whatsoever. (2006a: 42)[35]

Cantor's merit is to have offered mathematics a new method of treating pure multiplicities (and therefore of correctly doing ontology: Badiou admiringly refers to Cantor as 'the father of the great laws of our thought' (2008a: 45)), disseminating number into the multiplicity of sets. His theological commitments, however, made it impossible for him to embrace the radical un-grounding of inconsistent multiples in the void. He therefore attempted to save a place for God in his mathematical universe and posited, in the spirit of classical metaphysics, a divine absolute as infinite guarantor of Being, the (truly) infinite framework supporting his realm of the transfinite. It was Cantor's belief that all mathematics could describe was the (newly discovered) realm of the transfinite, but that it was up to speculative metaphysics to define the nature of the Absolute infinity of God.[36] This, Badiou argues, is the signature move of all ontologies of presence: 'the decision to declare that beyond the multiple, even in the metaphor of its inconsistent grandeur, the one is' (2006a: 42).[37]

It was only in the early decades of the twentieth century that Ernst Zermelo and Abraham Fraenkel developed an axiomatisation of Georg Cantor's original intuitions, in order to rid the theory of those paradoxes engendered by the notion of a totality of sets[38] which were discovered in the wake of Cantor's work: such a group of axioms (plus rules of derivation) was thought of as being able, in the spirit of the logicist project of the time, to account for the entirety of mathematics, all talk of mathematical objects being reducible to talk about sets.[39] The list of axioms enumerates the allowed operations with which, starting just with the empty set, the whole universe of sets *V* can be constructed. Stated informally these axioms (the Zermelo–Fraenkel axioms plus the axiom of choice (ZFC)) are:

0. Axiom of set existence or the void set: there is a set (and this is the void set).
1. Axiom of Extensionality: if two sets have the same elements (members), they are equal;
2. Axiom of the Power Set: for every set *x* there is a set *P*(*x*) whose elements are all the subsets of *x*. (a Subset being any set *y* which is part of the set *x* that groups together some elements of *x*);
3. Axiom of Infinity: there is an infinite set;
4. Axiom of Replacement: if the set *x* exists, the set *y*, obtained by replacing one by one each element of *x* with another element, also exists;
5. Axiom of Union: for any set *x*, there is a set *y* whose elements are the elements of the elements of *x;*
6. Axiom of Separation: for any set *x* and any well-defined property *a* there is a set *z* which is the set of elements of *x* which bear the property *a;*
7. Axiom of Foundation: any nonempty set *x* has at least one element *y* whose intersection with *x* is void (that is, no element belongs to both *y* and *x*);
8. Axiom of Choice: for a set *x* composed by nonempty elements which do not have any elements in common, there is a set *y* that contains exactly one element from each set in *x*.

What is most conspicuous in this list is that ZFC axiomatisation contemplates only two *existential* axioms; that is, only two axioms (in the list, 0 and 3), instead of describing self-evident properties of sets, postulate the *existence* of some sets to begin with. This can be expressed by the fact that ZFC axiomatisation employs first-order logic wherein the *existential* quantifier (∃, 'there exist') figures only twice in the list of axioms, to establish the existence of the empty set and of the infinite set, while all other axioms employ the *universal* quantifier (∀, 'for all'). Herein, for Badiou, lies the unique conceptual power of set theory, for what it *avoids* doing is to offer any *definition* of a set. The most primordial statement of any set-theoretical approach is the simple existential statement 'there is a set' or 'let X be a set' (axiom 0 above). Lacking a definition, how can a set be recognised? A set is that abstract object to which other objects *belong*. The relationship of belonging, indicated by the symbol ∈, is all that is necessary to discern a set: a collection of objects *belongs to* a set. If a ∈ x, b ∈ x and c ∈ x, the set x = {a, b, c}. It is already easy enough to understand how this intuition supports (of course, only if the mathematics = ontology equation is assumed) Badiou's rejection of ontologies of unity, and its demotion of the One to a constituted (counted-as-one) one-ness: 'the One is not' means that it (a set or, equivalently, a situation) has no definition/independent existence but that it is identified merely as the resulting product of the operator of belonging, of a count-as-one which lists all those elements which belong to it.[40] Denying any unity to objects beyond a purely formal one-ness, set theory implicitly posits that every*thing* is really multiple. On the other hand, the axiom of infinity states the existence of a 'limit ordinal' set that does not succeed any another set. This set, however, is but the first and smallest infinite set from which, by the successor operation, follows an infinite series of transfinite ordinals, and its existence remains submitted to the iterative priority of the empty set, since 'the original point of being for the limit ordinal is the void and its elements are solely combinations of the void with itself' (2006a: 157).

The achievement of this purely formal, empty definition of a set, only positing a Being without-presence, was an accomplishment which did not follow immediately from Cantor's work. He claimed paternity of a '*Mengenlehre* [doctrine of sets]' and specified that '[b]y a 'manifold' or 'set' [*Menge*] I understand in general any many [*Viele*] which can be thought of as one [*Eines*], that is, every totality of definite elements which can be united to a whole through a law' (Quoted in Hallett 1984: 33) and again that '[b]y 'aggregate' [*Menge*] we are to understand any collection into a whole [*Zusammenfassung su einem Ganzen*] M of definite and separate objects m of our intuition or our thought. These objects are called the "elements" of M' (Cantor 1955: 85).

Cantor built his theory on an (intuitive) understanding of a set as a collection of 'definite elements' of 'our intuition': Badiou wants to reject this vision.[41] ZF axiomatisation permits us to construct the entire universe of numbers from the empty set alone, and sets need never be defined. The only tool in the construction of the set-theoretic universe is the operator of belonging. No reference to objects of perception or consciousness is necessary, nor does any *intrinsic* property differentiate 'sets' from 'sets of sets', to the extent that these are only defined *relationally*. So, for example, the iterative construction of Von Neumann ordinals – where each ordinal is the well-ordered set of all smaller ordinals which precede it – can easily be performed with only one inaugural existential commitment: the empty set, negatively defined in terms of that set to which nothing *belongs*. The algorithmic process of construction of a mathematical universe starts with *nothing*, not with some ur-element given to intuition. Badiou assigns profound philosophical significance to this insight. All philosophical thought develops from an affirmation of nothingness towards infinity. Such was the (arch-rationalist) founding gesture of Descartes (the philosopher after whom, second only to Plato, Badiou fashions much of his style and method): to begin with the nothingness left behind by hyperbolic doubt and employ the mere *affirmation* of this nothingness as cornerstone of an entire philosophical edifice. Unlike Descartes, however, Badiou (through set theory) is able to turn the *subjective* founding gesture into an *impersonal* (ontological), axiomatic inscription of the Void as the name (or the *trace*) of the 'originary' nothing or inconsistent multiplicity. So the symbol Ø is to be understood as an arbitrary, indifferent signifier of the no-thingness devoid of any determination; indeed, Badiou suggests, the inconsistent multiplicity of Being is *there* in the very meaninglessness of its inaugural scriptural inscription.

In sum, the Zermelian axiomatization of set theory, in Badiou's reading, allows us to dispense with two theological remnants in Cantor's *Mengenlehre*: the divine nature of absolute infinity and the understanding of a set as a collection of units. Both these assumptions betray a subservience to the One as either constitutive basic *unit* (*qua* defined entity) or all-encompassing absolute *unity* (*qua* ultimate horizon of the mathematical universe). An axiomatic break with any kind of constructivist method – since '[d]efinition is the linguistic way of establishing the predominance of the entity' (2006b: 45) – or onto-theological thought is needed, and set theory stipulates precisely this break when laying down its two basic existential commitments:

> just like the empty set, or zero, the infinite will not be deduced: we have to decide its existence axiomatically, which comes down to admitting that one takes this existence, not for a construction of thought, but for a fact of Being. (2008a: 44)

We are now in a position to understand fully the link between set theory and atheism. In Badiou's view:

> mathematics localizes *a plurality of infinites* in the indifference of the pure multiple. It has processed the actual infinite via the banality

> of cardinal number. It has neutralized and completely deconsecrated the infinite, subtracting it from the metaphorical register of the tendency, the horizon, becoming. It has torn it from the realm of the One in order to disseminate it – whether as infinitely small or infinitely large – in the aura-free typology of multiplicities. By initiating a thinking in which the infinite is irrevocably separated from every instance of the One, mathematics has, in its own domain, successfully consummated the death of God. (2006b: 38)

The axioms of set theory are for Badiou the axioms of ontology itself, thoroughly de-sacralised, post-metaphysical and post-poetical. Ontology is the 'presentation of presentation' (2006a: 27) insofar as, ruled by these axioms, it is a situation that does *not* present (or count-as-one) Being itself, but presents what can *in principle* be presented about it, without reference to any ontic sphere of content, and without predicating anything about it other than its (inconsistent) multiplicity.

1.4 The Void

Set theory's most crucial axiom, in Badiou's reading, is that declaring the existence of the empty set. I have already mentioned that for Badiou all that *there is* are presented multiplicities, or *situations*. Indeed there is '*nothing* apart from situations' (2006a: 25: emphasis added). This can be read as an equivocal statement since inconsistent multiplicity *is* nothing insofar as it is not presented. And, in these terms, situations are *made of* no-thing. Situations are presented multiplicities made consistent by a count or structure. But multiplicities of what? We have seen that the inconsistent multiplicity that precedes structured situations cannot strictly be thought as the field of 'the many' (a determined multiplicity).

If the basic building block cannot be a 'One', what is it? The answer is, of course, a multiple, but a multiple *of nothing*: the empty set or the Void. Every purported whole is grounded on a deficiency, on a non-whole. The Void is the link between the situation and that vanishing 'stuff' from which the situation has been counted, the connection (or *suture*) between what *is* (*qua*) presented and what is presented (where *sous rature* means that the 'stuff' of presentation *is not* before it *is* presented). The Void, then, represents the proper *name* of Being, identified with the symbol $\emptyset$, and mathematics is the science of the productive manipulation of such Void. When included in a presentation (and *only* then) the proper name of the nothing of inconsistent multiplicity is the Void, *from which everything has always already originated.* Badiou puts it clearly: 'every thinkable being is drawn from operations first applied to the void alone. A multiple will be all the more complex the longer the operational chain which, on the basis of the void, leads to its determination' (2009: 112).

From within the ontological situation, the primordial act is the assignment of the proper[42] name 'Void' (and the symbol Ø) to nothingness, the paradoxical name where Being (the positive assertion of the existence of the empty set) and nothingness (which this name signifies) are indistinct – indeed 'there is a certain sense in which it alone [the Void] "is"' (2006b: 60). As such, the Void is the beginning of all philosophical (and mathematical) thought, which ultimately amounts to no-thing more than an iterative repetition (according to the successor operation) of this originary name.[43] Badiou writes:

> '[z]ero exists' is inevitably a first assertion; the very one that fixes an existence from which all others will proceed...Number comes first here: it is that point of being upon which the exercise of the concept depends. Number, as number of nothing, or zero, sutures every text[44] to its latent being. The void is not a production of thought, because it is from its existence that thought proceeds. (2008a: 22, 23)

In every situation, the Void is the name of the nothing that cannot be presented but that has logical priority over presentation itself. Why is that so? Because, lacking any predicable quality, any discernible feature, the only way for ontology to refer to the no-thing of inconsistent multiplicity is via an (axiomatic) naming.[45] The void is the intra-situational name of no-thing and is expressed, mathematically, by the empty set, the *only* set to be a multiple *of nothing*. Badiou explains:

> the name of the void is a pure proper name, which indicates itself, which does not bestow any index of difference within what it refers to, and which auto-declares itself in the form of the multiple, despite there being nothing which is numbered by it. (2006a: 59)

This void, ubiquitously found in any situation as the 'suture to its being' (2006a: 55) will allow Badiou to delineate the possibility of radical, extra-situational (and extra-ontological) novelty in his theory of evental truth, which I examine in Chapter 5. The Being of inconsistent multiplicity can only be summoned through a mere act of nomination, never directly examined in its non-structure. It is present*ed* but never present, eluding recapture by the One. To this extent, Badiou's is a *subtractive ontology*, where 'the rigour of the subtractive' is opposed to 'the temptation of presence' (2006a: 27).

1.5 Secularising infinity

Given the Void, an infinity of infinities follows. This is Badiou's vision of the ontological situation, regulated by ZFC set theory. But this insight into the nature of infinity was a hard-won historical achievement. Tracing the history of the relationship between infinity and finitude in Western thought, Badiou singles out three broad historical sequences.[46]

The classical, Greco-Roman pre-Christian sequence has a purely negative assessment of infinity as the *a-peiron* or absence of limit. This infinity lacks a proper existence, since the perfection of Being is considered to be dependent on the existence of limits – to exist is to have a limit, and the lack of limits characterises an indeterminate non-reality.[47] 'For a Greek', writes Badiou, 'what-presents-itself affirms its Being within the strict disposition of a limit (πέρας)' (2006a: 74). The ordered, harmonious *cosmos* is thus a finite universe, and the human ambition of transcending limits is a *hubris* that inevitably leads to tragedy, human values being reciprocal of the values of an ordered nature. The Greek Absolute was the regulative ideal of a perfect, *finite One*, where Being and existence totally coincide: for the Greek mind, the finite is the norm, the infinite a pathology.

The second sequence is the Christian-Romantic (Christian in its origin, Romantic in its final historical articulation) fusion of the One and the infinite under the name of God. Judaeo-Christian monotheism enforces the philosophical concept of an infinite perfection in the uniqueness of God. God is seen as the infinite creator transcending a finite world without direct commensurability between the finite human being and the infinite deity (unlike the classical world).[48] Yet, Christian monotheism is not a complete break with finitism: in yet another Heideggerian diagnosis Badiou writes that 'in the metaphysical age of thought, which fuses the question of being with that of the supreme being [onto-theology], the infinity of the God-being can be based on a thinking in which being, qua being, remains essentially finite' (2006a: 142). This second sequence opens up the thought of infinity only insofar the latter is seen as a totality or whole. So, the gap between God and beings engenders a new figure, proper to the Christian age, that of a *mediator*, whose paradigm is that of Christ: the human/divine mediator between infinite God and finite creation, between the unity of God and the multiplicity of creation. Philosophically, this originates dialectical thinking as a schema of *mediation* between contradictory terms, affording a grasp of the Absolute as the founding gesture of (Hegelian) romanticism. The Romantic stance is that of a finite creature striving to access the Absolute infinity of God through a creative effort capable of realizing the originary dialectical co-implication of the finite and the infinite.[49] Badiou writes that '[s] omething...is determinate for Hegel only insofar as it can be thought as other than another.... Determinateness comes down to the following: in order to found the Same it is necessary that there be some Other within the other. Infinity originates therein' (2006a: 162). For the 'methodical thinker of the Whole' (2009a: 142), then, infinity can always be tamed by (or extracted from) the self-comprehending activity of a totalised Absolute Being.

The third sequence is modernity, beginning with the scientific revolution spawned by Galileo, whose laws of motion, developed through an anti-Aristotelian mathematised physics,[50] introduced the concept of an infinite universe, and which culminates in Cantor's work (a sequence originating in a *scientific* revolution of thought as opposed to the previous one, originating in the *theological* union of Greek and Jewish thought). Not only God, but the world itself, in its multiplicity, was now considered as infinite.[51] Badiou explains that

> [t]he audacity of the moderns certainly did not reside in their introduction of the concept of infinity, for the latter had long since been adapted to Greek thought by the Judeo-Christian foundation. Their audacity lay in ex-centring the use of this concept, in redirecting it from its function of distributing the regions of being in totality towards a characterization of beings-qua-beings: nature, the moderns said, is infinite. (2006a: 143)

Yet, to declare the infinity of the natural world is not enough to decouple infinity and the One, and indeed could be re-assimilated as 'a gesture of presence which guarantees that the One is' (2006a: 144).[52] What is necessary is an (axiomatic) statement of the infinity of Being itself (and of the inexistence of the One), a declaration concerning 'the pure multiple, which is to say presentation' (2006a: 144). Again, Badiou refers to the ontological difference: it is not enough to observe an ontic infinity of natural entities; it is necessary to state the ontological infinity of pure presented Being. And this ontological infinity can *only* be properly said mathematically. So for Badiou

> [i]t is evident to what degree Cantor's oeuvre completes and accomplishes the historical Galilean gesture: there at the very point where, in Greek and then Greco-Christian thought, an essential appropriation of being as finite was based – infinity being the ontic attribute of the divine difference – it is on the contrary of being as such and of it alone that infinity is from this point on predicated, in the form of the notion of an 'infinite set', and it is the finite which serves to think the empirical or intrasituational differences which concern beings. (2006a: 145)

Only after the Cantorian revolution does the One wane in the West, a revolution which was brought about by an axiomatic decision – since no amount of empirical (ontical) observation could have broken the 'mechanism of ontological finitism' (2006a: 148) – declaring the existence of a mathematically rigorous infinity of infinities.[53]

Badiou outlines a number of possible consequences of this axiomatic declaration *vis-à-vis* theology. First, one could try to seek a new definition of God, not reducible to its attribute of infinity – which now can be attributed to the world itself. This is the way of those constructing a theology of a *weak* God, forfeiting the patriarchal traits of the Old Testament vengeful Father-God for an evental conception of God.[54] On the other hand, the focus on the infinity of nature is what engenders contemporary forms of neo-paganism or Heidegger-inspired 'deep ecology', preaching a return to nature as something sacred and divine in itself.[55] Second, one could find a new concept of infinity, or better, multiply the meanings of 'the infinite' in order to reintroduce a supreme space for the Divine. This is Cantor's attempt to place the absolutely or 'proper' (*Eigentlich*) infinite beyond the hierarchical proliferation of mathematical infinities. Third, one could declare the death of God and fill its vacant place with the Human. Affirming the infinity of humanity itself is what modern scientific humanism has attempted from Comte to contemporary 'new' atheism. Fourth, one could attempt to suppress infinity. This is the reactionary classicism grounded on a thinking of finitude (of the phenomenological-Heideggerian or positivistic-Wittgensteinian kind), which maintains infinity only as a regulative ideal, a field within which only finite operations can take place. Fifth, a return to an infinite One is possible, in the form of the affirmation of the infinite creative potentiality of life. This is the philosophy of the immanence of a virtual plane of forces that runs from Nietzsche to Deleuze through Bergson, rejecting a transcendent God while affirming the immanent vital drive of the world as such, in the form of a virtual reservoir of infinite potentialities to actualise. These forms of vitalism seek an immanent connection between the infinite and the One underlying the becoming of Nature.[56] Finally, the last option is to militantly pursue the severing of the infinite from the One, making it impossible for the latter to reappropriate the former: this is Badiou's project, which affirms the pure one-less infinity of the multiple as such.[57] This is not merely a new opposition between the finite and the infinite, but a deployment of an infinite multiplicity of infinities deriving from the mathematically articulated possibility of thinking real differences between infinities. This rational secularisation of the infinite,[58] then, does not just fulfil the death of God, but enforces a radically *deflationary* declaration of death of *any* metaphysical One.

In this chapter I have examined Badiou's ontology as a step in the right, immanentist and materialist, direction: first, axiomatically and 'subtractively' evacuating our metaphysics of self-present substances (indeed voiding 'matter' of any substantialist connotation) or transcendent grounds ('grounding' all there is in the groundlessness of inconsistency); second, pursuing the secularisation of infinity. Can we wholeheartedly accept Badiou's ontology? My answer is no, for it presents some major problems. As I will explain in the next chapter, what I take to be the major problem lies in its inability to offer a precise account of the relationship between the ontological and the empirical. How are non-ontological 'situations' to be understood? Is the common Badiouian exemplification of situations within one of the four fields of philosophical conditions to be understood as a mere analogy or is there something deeper than a structural similarity between, say, the 'situation' of the French Revolution or that of Galilean science and the ontological One? More precisely: how can we square Badiou's mathematical ontology with the philosophy (and the practice) of science whilst preserving (and endorsing) its de-theologising potentiality?

I take the two parts of this task to be complementary. It is only by naturalising Badiou's philosophy that we can make it truly impervious to theological thought, since several attempts have already been made to exploit Badiouian concepts to draw his system back to the theological fold. Consider, for example, Kenneth Reynhout's 'thought experiment' (2011: 220) to test the 'speculative thesis' (ibid.) that Badiou might be a 'hidden theologian' (borrowing the term from Paul Tillich) and that a post-onto-theological God could be identified with Badiou's Void (again through a Tillichian lens, favouring a 'God above God' or even a 'non-existent God' to the God of traditional theism). Reynhout holds that having surpassed an understanding of God as a supreme

Being among beings, theologians could aim at a creative appropriation of Badiou's ontology. The strategy would be to 'turn Cantor's absolute on its head' (2011: 231). Reynhout notes that even if rejecting Cantor's absolutely infinite inconsistent multiplicity (that is, Cantor's God) Badiou 'retains the distinction between consistent and inconsistent multiplies, relocating inconsistent multiplicity in being itself, the presentation of which is the void' (ibid.). This allows Reynhout to maintain 'Cantor's identification of God with inconsistency, but with Badiou relocating this inconsistency in being-itself and the presentation of the void' (ibid.). Frederiek Depoortere (2011) defends another possible compatibility between Badiou's ontology and theology, grounded on a critique of Badiou's (and Zermelo's) axiomatic rejection of absolute infinity via the axiom of foundation. Depoortere, therefore, rejects Reynhout's 'God = void' reading, but does so on exquisitely theological grounds: favouring a more orthodox and assertive, Thomist understanding of God's existence, he writes that 'the problem with Reynhout's proposal ... is the same problem as many contemporary versions of negative theology, namely that the *via remotionis* is uncoupled from the *via eminentiae* and that the latter is dropped so that only remotion is left' (2009: 123–4). For Depoortere to equate God with the Void means to 'fall in a trap that Aquinas warned us against, namely identifying God with the *esse formale*... [i]n this way, we run the risk of ending up with a complete identification of God and "being itself", without any distinction between Creator and creation' (2009: 124). To naturalise Badiou's ontology means to close these interpretative avenues, and Reynhout's reading should make us particularly aware of the necessity of questioning Badiou's separation between a presenting ontology and an unpresentable, inaccessible and inconsistent Being.[59] It is to the first step in a naturalising project of Badiou's philosophy that I will now turn.

2
The Ontological and the Empirical: Naturalist Objections

I concluded the previous chapter arguing that the relationship between the ontological and the empirical is the most intractable issue engendered by Badiou's system. Several of his commentators have identified this as precisely the main shortcoming of Badiou's philosophy. Feltham writes that

> in the absence of any verification of which sets correspond to which concrete situations, set-theory ontology cannot offer any rules concerning the existence of those entities we speak of in ordinary language. This incapacity is the source of one of the most serious charges laid at Badiou's door (from an empiricist quarter): his set-theory ontology is a castle built in the air or an 'ontology of a lost world'... At the very most, set-theory ontology declares that any identity claim whatsoever concerning personal identity, or distinct events inasmuch as it is based on a well-formed formula, must separate out its 'entity' from a larger presupposed multiple. Such is its materialism... But then what is its function if it is not going to explain the world or sort out other discourses and their existential commitments? (2008: 94)

Toscano similarly observes that '[r]ightly attacking the dogma that "science does not think" Badiou falls into the opposite trap, which he seems to gladly welcome, of Platonizing science into a radically useless, "immaterial" pursuit' (2000: 227). Hallward notes that 'by separating so decisively the ontological from the material or the physical, Badiou introduces a new dualism at the heart of his radically univocal arrangement' (2003: 276) whereas Osborne (2007: 24) objects that for Badiou,

> [o]ntology is severed from all phenomenological relations to objects ... only because Badiou decided to so sever it, in advance. Then he has the awkward task of restoring the connection between his set-theoretical mathematical entities, philosophically received ontological concepts (like nature and history) and the world. (2007: 24)

And, finally, Fraser flags the unanswered question of 'how one is to establish a correspondence between the sets presented in the ontological situation, and other, "concrete" presentations presented elsewhere' (in Badiou 2007a: xl). With this chapter I begin to propose a possible solution to this quandary.

Badiou is a rationalist *qua* Platonist: his commitment to the absolute power of reason (finding its highest expression in mathematics) finds warrant in the primacy of the intelligible over the sensible and in the equivalence of thought with Being. In Badiou's reading,

> Plato's fundamental concern is to declare the immanent identity, the co-belonging, of the knowing mind and the known, their essential ontological commensurability. If there is a sense in which he remains an heir to Parmenides, who declared 'it is the same to think and to be', it is to be found in this declaration. (2006b: 52)

Badiou's 'noocentrism'[1] will appear problematic to the realist committed to what Bhaskar (2008: 17) calls the *intransitive dimension* of external reality: reality, and the mechanisms according to which it operates, must be independent from our thought of it. Between thought and Being, to borrow the Laurellian term, there is a *unilateral duality*. Badiou pre-emptively defends himself from the charge of idealism by floridly describing his election of mathematics as rational discourse of being as in line with a 'glacial anti-humanism', allowing for a 'trans-human advent of truths', mathematics being 'an instance of stellar and warlike inhumanity' (Badiou 2006b: 14). He goes as far as describing his own system, with considerable bravado, as 'the most rigorously materialist in ambition that we've seen since Lucretius' (Badiou 1994: 123). Ray Brassier has identified this discrepancy between the idealist (Platonic-Parmenidean) and the materialist (Lucretian) Badiou, and explained that their reconciliation

> is deceptively simple. It is the identification of being as void through axiomatic set theory that purges materialism of the methodological idealism whereby matter is reinscribed in a concept. By embracing a subtractive ontology, materialism requires only one name for being:

> that of the void or null-set, Ø. Being and thinking are 'the same' to the extent Badiou tends to define both of them subtractively – as void and truth respectively. Being is void as subtracted from presentation, while thought is truth as subtracted from knowledge. (In Hallward 2004: 51)

Brassier correctly notes how the identification of Being (and thought) with pre-presentational nothingness is not warranted by any intra-mathematical axiom but by a purely philosophical decision. However, granting consistency to Badiou's identification of thought and Being within the confines of his system still does not satisfy the philosopher interested in squaring her ontological project with the practice of modern science. Badiou's rejection of the 'ideological split' between matter and thought (or content and form) takes a rationalist high road which emphasises the productive, materialist character of mathematical thought *qua* inscription at the expense of any mooring in *any* of the natural sciences. A solution to Badiouian noocentrism can be identified by (1) restating his equation of thought and Being (meant to render 'immanence and transcendence indiscernible' (2006b: 52)) as a *naturalist* thesis, indicating that the transcendental conditions of thought lie in the material, immanent substratum of natural processes which instantiates minds (the real is the objective condition of subjective thought); and (2) defusing the matter/mathematics dualism by postulating mathematical objects (or more correctly, as I will show in Chapter 4, structures) as all that there ultimately is to material reality. The materialist aim is to explicate an intransitive dependence of thought (and that which does the thinking) upon that which is thought: the former being prior in the logical order of understanding, while the latter enjoying ontological priority in the order of explanation.

2.1 Pythagoreanism and materialism

I endeavour to move between empiricism and rationalism through a simultaneous commitment to (scientifically negotiated and conceptually mediated) sensibility as a source of knowledge of the external world, and to the capacity of human reason to reach beyond sensible perception in abductively discerning the hidden structures of reality: that which is observed in the natural sciences and that which is thought in mathematics share one and the same ontological status.

This stance is often dismissed by Badiou and his commentators as too untenable to deserve real consideration. To mention but a few instances:

Badiou writes that 'except if we Pythagorize, there is no cause to posit that being qua being is number' (2006a: 24); Hallward observes that 'actual beings are clearly not themselves numbers' and that 'Badiou's ontology is not a fanciful return to Pythagorean speculation' (2003: 63); Pluth assures us that we can 'rule out any sort of number mysticism on Badiou's part' since 'he is not a classical Platonist, and he does not think the nature of being *is* mathematical' (2010: 46); and, finally, Bartlett and Clemens (2012), in a highly polemical retort against some critics of Badiou, scathingly chastise the latter's misreading of Badiou as a Pythagorean – in their assessment too ridiculous a stance to be seriously contemplated. These responses, if correct from the point of view of Badioiuan exegesis, suffer from two flaws. First, if there is any consensus among historians it is to the effect that there is no firm ground to affirm anything certain regarding the details of Pythagorean doctrine, over and above its esoteric and exclusive nature, its interest in numbers and its concern with the transmigration of souls.[2] Thus to invoke the Pythagorean ghost is little more than a rhetorical sneer. Second, and more importantly, even admitting that we can understand the name 'Pythagoreanism' to indicate any philosophy giving metaphysical pre-eminence to number and mathematics, it is far from clear what 'number and mathematics' actually *mean* in this context, as well as why such a metaphysical choice should be a form of 'mysticism'. It is my contention, against these ironic dismissals, that a suitably modified 'Pythagorean' option would be uniquely able, having clarified what 'mathematics' is about, to resolve Badiou's problems and to offer a thoroughly immanentist and demystified metaphysical worldview.

An uncrossable boundary between the domains of physics and mathematics has often been expounded by Badiou himself. Let me quote at length some rare passages where Badiou directly addresses the problem of the relationship between the two:

> [i]n the final analysis, physics, which is to say the theory of matter, is mathematical. It is mathematical because, as the theory of the most objectified strata of the presented as such, it necessarily catches hold of being-as-being through its mathematicity. The relation between 'what is presented', for example, matter, and the theory of being as being can be described, empirically, as the relation between physics and mathematics. But it might be described more profoundly, as the relation between, on the one hand, a generic theory of the multiple in itself – that is, of a multiple indifferent to what it is the multiple of, and thus of the multiple as pure multiple of the multiple – and,

> on the other hand, the 'that which is multiply presented as such', about which ontology says nothing... the more you decompose the concept of matter into its most elementary constituents, the more you move into a field of reality which can only be named or identified with increasingly complex mathematical operations. 'Matter' would simply be, immediately after being, the most general possible name of the presented (of 'what is presented'). Being-as-being would be that point of indistinction between the possible and the real that only mathematics apprehends in the exploration of the general configurations of the purely multiple. Matter, in the sense in which it is at stake in physics, is matter as enveloping any particular presentation – and I am a materialist in the sense that I think that any presentation is material. If we consider the word 'matter', the content of the word 'matter', matter comes immediately after being. It is the degree of generality immediately co-present to ontology. The physical situation will then be a very powerfully mathematised situation and, in a certain sense, more and more so, the closer it comes to apprehending the smallest, most primordial elements of reality. (1998: 127)

Or again:

> [m]athematics is a necessary formal dimension of all scientific discourse, if we understand by 'science'... the rational theory of those phenomena in the world which do not depend directly upon the conscious activity of man... The mathematical exigency is formal, in so far as it supports, as to the intelligibility of these phenomena, their most abstract and most general strata, that which relates to their pure being, to their multiple composition. But, precisely, this strata cannot, in my view, represent that which is strongest and most 'true' in those sciences which are not purely mathematical. A crucial point about physics is to present, to create concepts, so that they can be mathematically expressible, all the while retaining a relation to the world which means that they cannot be deduced from any mathematical corpus whatsoever. This is the case with the concept of uniform movement in the principle of inertia. Moreover, it is this irreducible worldly dimension which opens onto the possibility of experience, at the same time as mathematical formalisation guarantees the universality of experimental results, in the form of their always-possible repetition... So: I neither believe that physics is 'reducible' to mathematics, nor do I believe that mathematics 'founds' physics. Between the two, there is a rooting of concepts in a determinate

> world, which the experimental method designates and delimits, in a gesture which is of a transcendental nature... we can say that there exist two sciences which cannot be hierarchised: ontology, or pure mathematics, and physics, the science of those worlds accessible to our experience. (2007b: 17, 18, 19)

These passages clearly display unresolved tensions in Badiou's thought, ultimately caused by the uncompromisingly rationalist election of mathematics as queen of the sciences, which makes it impossible to place it on a continuum with other natural sciences. Indeed, this unilateral prelation of mathematics seems inconsistent with Badiou's own recognition of something like a 'Galileo-event' (Badiou 2005a: 38) as that historical moment where the truth of the indifference between physical reality and mathematical description came to be recognised, thus triggering the creation of modern mathematical physics.[3] The separation of physics from (or irreducibility to) mathematics depends on Badiou's (all too Heideggerian) unwillingness to do away with the split between the 'indifferent multiple thought as such' or Being and the world of physical matter between pure presentation in thought and phenomenal presentation in external reality; between the ontological and the ontic. Given that indeed 'the more you decompose the concept of matter into its most elementary constituents, the more you move into a field of reality which can only be named or identified with increasingly complex mathematical operations' (a statement which would be conceded by any physicist), it seems hard to identify *where* the cut-off point between matter and Being would lie once we question the explanatory value of the distinction between an inconsistent multiplicity of Being and presented consistent multiplicity. In other words, even the *subtractive* nature of inconsistent multiplicity reintroduces a form of the same ontological difference on which Heidegger grounded his critique of science. Badiou may paint himself into Heidegger's corner when he explains that he is 'a materialist in the sense that I think that any presentation is material' while at the same time claiming that 'matter comes immediately after Being'. Wouldn't a materialist want to claim that there *isn't* any 'Being' immediately before matter? And how are we precisely to cash out the metaphor that physics '*catches hold* of Being-qua-Being'?[4]

Galilean mathematisation was (as Badiou correctly observes) an intellectual procedure which, *starting with* observation and measurement of a given phenomenon, sought to formulate a phenomenological law in mathematical terms, or to subsume a vast number of physical processes under a general mathematical law capable of representing the evolution of such processes.[5] Shortly after Galileo, Newton took an additional step in mathematical abstraction: by considering the phenomena and the laws describing them he attempted to derive *more general* mathematical laws (for example, the law of universal gravitation) capable of subsuming and explaining all of the single phenomenological laws (for example, those quantitatively describing the rate of fall of a body) – hence at a further remove from lived experience. The twentieth century brought with it yet *another* level of abstraction. Boniolo and Budinich refer to 'Dirac's methodological revolution' (in Boniolo et al. 2005: 91), referring to the change that the scientific work (and implicit philosophy) of Paul Dirac brought about in physics, wherein mathematical thought became 'an inductor of discoveries of *new phenomenic aspects* that enhance and develop our knowledge of the world' (ibid.: emphasis added). Arguably *the* conceptual revolution in twentieth-century physics is the almost complete severance between (in Sellarsian terms) the manifest image of the empirical world and the scientific image of the subatomic, mathematised reality studied by science. So, to refer again to Badiou's claims quoted above, it is quite simply *not true* that the mathematised concepts employed by contemporary physics retain 'a relation to the world which means that they cannot be deduced from any mathematical corpus whatsoever'. This misunderstanding is ultimately the product of Badiou's lack of engagement with the empirical sciences. Adrian Johnston puts it in the most unequivocal terms when he claims that Badiou's indifference towards the sciences 'runs a very high risk of allowing for a relapse into theosophy' (2008a: 39). Indeed, his indictment of poetical philosophising notwithstanding, Badiou's prioritisation of mathematics as presentation of a subtracted Being risks producing the same anti-naturalist blindness to the achievements of the empirical sciences, which in turn leaves open the possibility for a form of Neoplatonic, mathematical mysticism, at odds with any materialism.

If it is true that 'the [Galilean] event of the creation of modern mathematical physics opens a sequence of the thinking or understanding of Nature' (Badiou 2004: 181), to be a materialist within the fold of this sequence means more than being hazily committed to the materiality of 'presentation'. The materialist equates 'matter' with *all* that there is (against Badiou's claim that matter is a 'particular [domain] of presentation' (2006a: 127)). And if elementary matter can only be spoken of mathematically, then this mathematical structure *is* all that there is: materialism, if anything, is a thesis that rejects all ontological dualities. Perhaps the crucial point is that Badiou's insistence on a necessary 'consistency' of what is presented/physical reality is what leads him to the construction of a gap between what is and what appears. Here, therefore, I wholeheartedly agree with Adrian Johnston's plea for 'ontic impurity', aiming to challenge

> the very possibility of simultaneously being a materialist (as Badiou professes to be) and (as Badiou allegedly does) accepting a clear-cut distinction between the ontological and the ontic. Authentic materialism, especially a materialism with a relation to dialectical thinking, must resign itself to the messiness of a theoretical account of the instances and types of being conditioned by the empirical results of such ontic fields as the natural sciences. A materialist should be deeply suspicious of the cleanliness of any ontology of pure being in and of itself. (2008a: 28)

Johnston's incredulous tone is also justified when he asks '[h]ow on earth can one claim to be any stripe of materialist, let alone the most extreme and consequent materialist in the two thousand years since Lucretius, if one equates being as such with intangibly abstract numerical structures ... ?' (2008a: 29). Yet this is where I part ways with Johnston, since I think that it is precisely through an equation of Being/matter with mathematical structures that we can remedy Badiou's materialist shortcomings, locating such structures not in an abstract mathematical (or noetic) space (a space of discourse), but identifying them with the real, physical world (a space of 'Being'). That is to say, Badiou's ontology can be naturalised through an encounter with structural realism. Undercutting Badiou's byzantine separation between Being (ontological situations) and appearing (worlds) in his later work,[6] such a solution will allow us to answer, as Johnston puts it elsewhere 'the unanswered questions of how and why being(s) give rise to worlds' (2008b: 349). In short, my naturalisation of Badiou proceeds by the *negation* of two core theses of his work (1) '[mathematics is ontology] is not a thesis about the world but about discourse' (2006a: 8) and (2) 'there is no structure of being' (2006a: 26). My wager is that most of Badiou's insights concerning the coming-into-being of set-theoretically describable structures and the 'production' of novelty can be salvaged once we operate a naturalist simplification of his system.

2.2 Philosophy of mathematics to the rescue

Badiou's philosophical approach to mathematics is radically distinct from the (analytic) tradition of the philosophy of mathematics, an area to which he assigns the unflattering sophistic label of 'the little style'

(2006b: 3). Such is the style of those interested in treating mathematics as bounded within logico-linguistic constraints, hence ignoring its ontological significance. For Badiou, this philosophy of mathematics is just 'a linguistic version of Aristotelianism' (2006c: 50), refusing to grant ontological significance to mathematics by stressing that mathematical objects are useful fictions. In this view '[m]athematics is...ultimately a rigorous esthetics. It tells us nothing of real-being' (2006c: 48), since it boils down to a rule-bound procedural explication with no ontological depth. What is necessary today is 'a Platonist rectification' (2006c: 50) of the relationship between mathematics and philosophy, one reminding us that 'in mathematical thought, or in mathematics as a thought, it is the real, and not mere words, which is at stake' (2006b: 173).

Indeed, Badiou's philosophy is, to use Ian Hacking's term, thoroughly *infected* by mathematics[7]: to the extent that the mathematical injunction conditions his whole thought, from ontology to politics, he can perhaps be compared (as he would want it to be) only to Plato. Badiou rejects any subordination of mathematics to philosophical jurisdiction: mathematics is a *condition* for philosophy, not vice versa. Only such an approach can properly be called the 'grand style', one which 'stipulates that mathematics provides a direct illumination of philosophy, rather than the opposite, and that this illumination is carried out through a forced or even violent intervention at the core of these issues' (2006b: 7–8). More bluntly put 'mathematics today has no need of philosophy' (2006a: 10). Mathematics cannot be domesticated as one philosophical field, via its reduction to logical or linguistic elements, but must be evaluated as independent. In Norris's reading,

> Badiou's major claim...is that philosophy of mathematics has sold mathematics grievously short by focusing on questions like: What is mathematical knowledge? How can we be certain that we have it? What can or must be the nature of mathematical entities such as numbers, sets, or classes if indeed we can have knowledge of them? It has thus been prevented from raising questions with regard to the primary (ontological as opposed to epistemological) issue of truth as that which might always surpass – and perhaps, in consequence of some future advance, eventually be known to have surpassed – a given, temporally indexed state of knowledge or present-best belief concerning it. (2011: 11)

However, Badiou's stance is far removed from a mathematical naturalism of the kind defended, most vocally, by Penelope Maddy.[8] If it is true that, for Badiou, mathematical practice needs no extra-mathematical justification or defence, it is also true that philosophy enjoys a meta-mathematical (that is meta-ontological) role which Maddy would not endorse. Only *philosophy*, Badiou claims, can fully assess the ontological insights (truths) unwittingly[9] disclosed by mathematicians. For Badiou, within the confines of the little style 'what is at stake is not what mathematics might entail for ontology, but rather the specific ontology of mathematics' (2006b: 4). Even Maddy's post-1997 naturalist scepticism towards ontological realism in mathematics would not exonerate her, in Badiou's eyes, from being a practitioner of the little style, since by no means is she interested in drawing ontological *conclusions* from mathematical premises (in fact rejecting the necessity of any ontology accompanying mathematical practice), but is mostly concerned with the empirical methodology of science.[10]

Badiou's clarion call for a return to Platonism rejects the label of 'Platonism' as commonly employed in the philosophy of mathematics and analytic philosophy in general, what in Badiou's eyes amounts indeed to an *anti-Platonism*. Orthodox mathematical Platonism is generally defined as that position which claims mathematics involves the study of independently existent mathematical objects and the discovery of verification-transcendent mathematical truths.[11] If Maddy (1990: 20–1) acknowledges that 'as it is common with such venerable terms, [the label "Platonism"] is applied to views of very different sorts, most of them not particularly Platonic', Badiou considers the 'Platonist' attribution of this position to be downright incorrect, since 'it presupposes ... a distinction between internal and external, knowing subject and known "object" ... which is utterly foreign to the genuine Platonic framework' (2006b: 51).[12] For Badiou '[t]he great problem of Platonism is not really the distinction between the intelligible and the sensible, but the understanding that sensible things participate in the intelligible' (2007a: 92). The 'realist and obsolete vision of mathematical objects' (1999: 121) attributed to Plato must be abandoned and this wrongheaded reduction of Plato's position to an -ism is nothing but the consequence of the 'little style' characteristic of Anglo-American philosophy, which

> assumes that mathematics can be treated as a particular area of philosophical concern; that this treatment necessarily proceeds through a consideration of logic and language; that it is entirely compatible with ready-made philosophical categories; and that it leads to a classification of doctrines in terms of proper names. (2006b: 5)

This, in Badiou's view, is nothing but narrow-minded scholasticism.[13] As my reconstruction of Badiouian ontology in Chapter 1 has established, Badiou can reject at once (or move in between) mathematical Platonism and formalism,[14] deflating the epistemological problem of access to mathematical entities and re-casting in ontological terms, by claiming that '[t]he truth is that *there are no* mathematical objects. Strictly speaking, mathematics *presents nothing*, without constituting for all that an empty game... not having anything to present, besides presentation itself' (2006a: 7).

Badiou's aristocratic and indiscriminate dismissal of the entirety of philosophy of mathematics is produced, as Ray Brassier observes, by his 'materialist critique of what he regards as the ideological distinction between "real" and "ideal", common to empiricism and idealism' so that his Platonism, as I have observed, would 'recuse the empiricist distinction between thought and object' (2005: 136). Badiou's goal is praiseworthy, his method less so. As I have explained above, there are two ways to reject (or deconstruct) the formal/concrete (or ideal/real) binary, one taking the rationalist high road (Badiou) and another which, in naturalist spirit, assigns ontological precedence to the real and prefers to start with scientifically delivered empirical data. An offhand rejection of the insights of the philosophy of mathematics is wrongheaded. First, the concerns of contemporary philosophy of mathematics are by no means an arbitrary selection operated by narrow-minded post-'linguistic turn' analytic philosophers but are (in direct line of descent from) the very issues that troubled first-rate mathematicians like Henri Poincaré, David Hilbert, Kurt Gödel, Hermann Weyl and John von Neumann.[15] Second, Badiou's claim that '[n]owadays there is an incontestable supremacy of the constructivist, indeed intuitionist vision' (2006c: 49) is, well, contestable, since it ignores realist developments both in the direction of a set-theoretic realism and of mathematical structuralism. Third, why should we accept Badiou's 'noocentric' understanding of Platonism as prescriptive? Even admitting the historical accuracy of such an interpretation, our philosophical concern with mathematics should be aimed at uncovering *how things are* (whether in terms of mind-independent realities or in terms of the cognitive basis for mathematical thought) rather than be subject to Plato's approval. Fourth, and perhaps most importantly, a significant slice of the philosophy of mathematics has dealt with the important problem of accounting for the relationship between mathematics and the natural world, a problem swept off the table *a priori* by Badiou's rationalist equation of Being with thought, but one that, as I have argued above, rears its head as the main obstacle for a concrete application of his philosophy to any non-ontological field. Ultimately, we can still draw resources from the tradition of philosophy of mathematics, without necessarily accepting the 'destitution of ontology' (2006c: 110) which followed the linguistic turn. In particular this fusion can help us with naturalising Badiou's insights into the ontological (what we can say about what there is) and epistemological (how we can acquire novel knowledge about it) significance of mathematics, according to the immanentist and naturalist imperative of placing them on a continuum with the rest of our theories about reality.

The issue of the applicability of mathematics to non-mathematical situations is a problem for both Badiou (as the pertinence of ontology for ontic situation) and the philosophy of mathematics[16] whether or not Badiou agrees. I have already exposed the problem from the Badiouian perspective, so it remains to be seen how this has been tackled – although not resolved – in the philosophy of mathematics. What is required is precisely the kind of synoptic view which I try to adopt in this book, and an open-minded – perhaps speculative – interdisciplinary programme.

2.3 The applicability of mathematics

The *locus classicus* from which these discussions derive is Eugene Wigner's landmark essay 'The Unreasonable Effectiveness of Mathematics in the Natural Sciences' (1960), wherein he mused on the surprising power of mathematical formalism to describe the natural world, concluding with the now-famous maxim: '[t]he miracle of the appropriateness of the language of mathematics for the formulation of the laws of physics is a wonderful gift which we neither understand nor deserve' (1960: 14). For Wigner, the fact that mathematical concepts created by mathematicians (following aesthetic criteria such as simplicity or elegance) could be successfully applied (in the formulation of 'laws of nature') to the theoretical description of empirical phenomena is to be accepted as an 'empirical law of epistemology' (1960: 10) for which we can find no explanation (Wigner does not shy away from talk of a 'miracle' and 'article of faith'). In recent years 'Wigner's problem' has come to be re-examined by a (small)[17] number of philosophers of mathematics.[18] The most thorough engagement with it has been Mark Steiner's, in a number of papers (Steiner 1989; Steiner in Shapiro 2005) and a monograph on the topic (Steiner 1998). Steiner (perhaps with excessive grandeur) claims that '[t]o an unappreciated degree, the history of Western philosophy is the history of attempts to understand why mathematics is applicable to Nature, despite apparently good reasons to believe that it should not be' (in Shapiro 2005: 625). He charges Wigner's essay with conflating heterogeneous problems, each depending on a different interpretation of the term 'applicability' and holds that the multitude of cases where mathematical concepts are applied to physical contexts should be approached one by one in piecemeal fashion. Moreover, Steiner remarks that 'Wigner's puzzle' should be seen as referring to both descriptive applicability and (more interestingly, in his view) to the discovery of new laws and phenomena: on the one hand, empirical regularities can be handled via a mathematical formalism that allows for the prediction of the occurrence of a given phenomenon (say, Mars' future position in the sky) given certain initial conditions (say, the known arrangement and position of the planets in the solar system);[19] on the other, mathematical concepts can facilitate the discovery of hitherto unknown phenomena and regularities. Steiner considers the former to be an uninteresting, logico-semantic phenomenon, but takes the latter as constituting a broader issue, one which 'concerns the applicability of mathematics as such, not of this or that concept', and that it is 'therefore an epistemological question, about the relation between Mind and the Cosmos' (1998: 45), that is, the 'actual' problem of unreasonable effectiveness, which Wigner only confusedly hinted at.

Steiner wants to reconceive the applicability problem in terms of the necessity of accounting for the success of the global strategy (1998: 5) adopted by twentieth-century physicists to discover features of the natural world. His thesis is that scientists arrived at the predictive formulation of laws for the subatomic nature of matter via forms of mathematical analogy, and that indeed they had 'no real alternative' (1998: 3). Steiner describes two kinds of analogies: Pythagorean and formalist. Pythagorean analogies[20] relying on similarities definable only in terms of mathematics, without reference to any physical system, so that 'any analogy among structures the mathematician did, or could, recognise, became a potentially physical analogy too' (1998: 4). So, in the example of Maxwell's discovery of the equations describing electromagnetic radiation[21] Steiner writes that Maxwell's reasoning was Pythagorean, since '[o]nce he had a mathematical structure which described many different phenomena of electricity and magnetism, the mathematical structure itself, rather than anything underlying it, defined the analogy between the different phenomena' (1998: 79).[22]

Formalist analogies,[23] on the other hand, are analogies where mathematical notation itself (the syntactic structure of mathematical symbolism) provided an analogy which led to physical discovery. In such cases 'the analogy was to the form of an equation, not to its mathematical meaning' (1998: 4).[24] In these cases the mathematical work was even further removed from the physical world than in Pythagorean analogies: indeed Steiner maintains that 'much physical research in the present century has been, in the first instance, inquiry into our own formalisms, only secondarily into nature' (1998: 116). What does the empirical success of these mathematical analogies tell us? How to justify the 'Galilean point' that 'in formulating conjectures the working physicist is gripped by the conviction ... that the ultimate language of the universe is that of the mathematician'? (1998: 5). Steiner is primarily interested in the implications of this phenomenon 'for our view of the universe and the place in it of the human mind' (1998: 2), and intends to draw more radical consequences from it than a conceptual or metaphysical Pythagoreanism (respectively, the thesis the essences of entities are mathematical structures and their relations and the thesis that such entities are nothing but mathematical structures), positions which he nonetheless considers 'very respectable' (1998:5)).[25] Mathematics, for Steiner, is an anthropocentric concept and, by extension, to employ mathematical analogies is an anthropocentric strategy. His 'major claim' is that 'relying on mathematics in guessing the laws of nature is relying on human standards of beauty and convenience' (1998: 7). The historical fact that mathematics has had predictive power in the empirical sciences thus leads to the conclusion that the universe itself is 'user-friendly' (1998: 8). Steiner's conclusion then is 'both that physicists acted as though they had (implicitly, for the most part) anthropocentric beliefs, and also that the world really does look anthropocentric – in the limited sense that it is intellectually accessible to human research' (1998: 9). This thesis, which Steiner describes as anti-naturalist, has theistic potential (which Steiner acknowledges; see 1998: 10) as it seems to require an unavoidable theistic justification: the thought that anthropocentric heuristic criteria of beauty and convenience are successful in tracking the objective structure of the universe implies that the universe itself was structured with these same criteria in mind by a creator. Like Steiner, I find the (metaphysical) Pythagorean solution respectable but, unlike him, I am not prepared to link intelligibility with anthropocentric design. Steiner's stance is indeed all too close to that of religious apologists. John Polkinghorne, for example, explicitly proposes a theistic answer to Wigner's Problem when he writes that

> I believe that Dirac and Einstein, in making their great discoveries, were participating in an encounter with the divine. ... The 'unreasonable effectiveness of mathematics' in uncovering the structure of the

> physical world...is a hint of the presence of the Creator, given to us creatures who are made in the divine image. I do not present this conclusion as a logical demonstration – we are in a realm of metaphysical discourse where such certainty is not available either to believer or unbeliever – but I do present it as a coherent and intellectually satisfying understanding. (1998: 4–5)

These interpretations are even shared by the official Vatican line on the matter. In a letter to Archbishop Rino Fisichella (then Rector of the Pontifical Lateran University) written on the occasion of the international congress 'From Galileo's Telescope to Evolutionary Cosmology: Science, Philosophy and Theology in Dialogue', (then) Cardinal Joseph Ratzinger announced that

> if nature is really structured with a mathematical language and mathematics invented by man can manage to understand it, this demonstrates something extraordinary. The objective structure of the universe and the intellectual structure of the human being coincide; the subjective reason and the objectified reason in nature are identical. In the end it is 'one' reason that links both and invites us to look to a unique creative Intelligence. (Ratzinger 2009: n.p.)

The metaphysical Pythagoreanism I want to defend need *not* imply, however, an orderly and intrinsically 'user-friendly' universe emanating from a transcendent cause. Paraphrasing Ratzinger, to conclude that nature is really mathematically structured and that mathematics is formulated by 'man' does not demonstrate anything extraordinary over and above the materialist persuasion that reality is immanently deployed in structural form. Badiou's insights are here expedient: mathematics has an immanent productivity and a mathematical ontology is grounded only on a referent-less and meaning-less empty set. Steiner is correct in articulating the question of the applicability of mathematics as involving a reflection on the universe and of our part in it, but his answer is opposed to mine, precisely to the extent that his answer is an anti-naturalist one (Steiner defines 'naturalist' a stance that rejects any anthropocentric point of view). Both the ontological question of how physical reality is amenable to mathematical description and the epistemological question of how we come to formulate mathematical concepts to begin with are to be asked. In order to answer the latter question, I will now survey a few important attempts to naturalise our mathematical concepts.

2.4 Naturalist explanations of mathematics

In her early work *Realism in Mathematics* (1990),[26] Penelope Maddy sketches what she defines as 'set theoretic realism' (1990: 35), a project that can be placed in the tradition of mathematical Platonism. However, her position diverges from orthodox Platonism insofar as she attempts to offer a realism capable of resisting the standard epistemological objections raised to it, famously voiced by Paul Benacerraf (1973). Briefly, Benacerraf mounts an attack (less fierce than it is usually represented as being) against the notion of a 'mathematical intuition', as mobilised by Kurt Gödel. Gödel, describing the mathematician's cognitive access to the mathematical realm of abstract (non spatio-temporal) objects, famously argued that

> despite their remoteness from sense experience, we do have something like a perception also of the objects of set theory, as is seen from the fact that the axioms force themselves upon us as being true. I don't see any reason why we should have less confidence in this kind of perception, i.e. in mathematical intuition, than in sense perception, which induces us to build up physical theories and to expect that future sense perceptions will agree with them, and, moreover, to believe that a question not decidable now has meaning and may be decided in the future. (1990: 268)

Benacerraf takes issue with this view, and holds that, in a (causal) account of knowledge acquisition, there must be a clearly identifiable *causal link* between knower and known, and it isn't clear how there can be any such link between abstract objects and the spatio-temporally located brains of mathematicians. The causal isolation of mathematical objects seems to foreclose the possibility of any epistemic access to them. Benacerraf notes that this kind of problem has been evident since Plato, and that Plato himself 'had recourse to the concept of *anamnesis* at least in part to explain how, given the nature of the forms as he depicted them, one could ever have knowledge of them' (1973: 675), implying – presumably – that such a solution is not acceptable in modern times. What became known as 'Benacerraf's dilemma', then, is this: given that we would like our referential use of language to operate in the same way in scientific and mathematical work alike, we should either resolve the issue of how ontologically suspect entities like mathematical objects can be referred to, or abandon, along with the semantic theory of direct reference, the causal account of knowledge,[27] or at least postulate some special, non-sensory way in which we get to access the recognition-transcendent objects that populate the mathematical realm.[28]

Maddy, having proclaimed her allegiance to scientific realism, defends a version of mathematical realism as a compromise between (the best of) Gödelian Platonism and the so-called Quine–Putnam indispensability thesis, the naturalist position of minimal metaphysical commitment which accords reality to mathematical entities only insofar as they play an indispensable role in scientific theories about the physical universe.[29] Maddy accepts the latter thesis, yet corrects the excessive restrictions imposed by it on mathematical justification with a Gödelian commitment to the independent existence of numbers stronger than the merely instrumental one assigned to them by Quine and Putnam, including a reliance on purely mathematical forms of justification (so to allow the reality of mathematical objects examined in pure mathematics, unapplied in physical sciences). However, in order to dodge Benacerraf's objections Maddy corrects (naturalises) Gödelian 'mathematical intuition' by anchoring knowledge of numbers in *empirical perception*. Her aim, she later summarised, is to replace 'Gödel's rather mystical faculty with a down-to-earth neurophysiological model that describes both physical and mathematical perception and uses these as the basis of both physical and mathematical intuition' (1997: 108).

Her proposal, then, is that sets are spatio-temporally located[30] and that numbers are nothing but properties of sets: numbers now have a causal interaction with the physical world (and our physical brains – thus solving the epistemological problem) and are not identified with sets but seen as properties thereof (solving the metaphysical problem). That is to say, when visually perceiving three stones on the ground, we are *also* perceiving a set with three objects, and when we mathematically refer to the entity known as 'set' we are generalising from known samples of physical objects. Such an empiricist-friendly realism could not be further from Badiou's (or Gödel's) Platonism.[31] Maddy argues that intuition is to be distinguished from perception and that the latter, giving rise to mathematical beliefs, is not qualitatively different from that on which we rely for our access to physical objects. She refers here to the pioneering work of neurophysiologist Donald Hebb on the formation of neuronal structures in learning and pattern recognition through the action of the so-called Hebb synapses[32] and argues that repeated encounters with certain physical arrangements of objects can modify the plastic neuronal structure of the brain, engendering cell assemblages which function as 'set-detectors', allowing us to see a certain pattern as an instance of a set.[33] This naturalistic thesis erases the difference between the natural scientist's perception of physical objects and the mathematician's perception of sets. Maddy explains that

> [w]hat goes for physical objects should also go for sets: the development of higher-order cell-assemblies responsive to particular sets gives rise to an even higher-order assembly corresponding to the general concept of set. The structure of this general set assembly is then responsible for various intuitive beliefs about sets, for example that they have number properties, that those number properties don't change when the elements are moved (barring mishap), that they have various subsets, that they can be combined, and so on. And these intuitions underlie the most basic axioms of our scientific theory of sets. (1990: 70)

The source of our beliefs about sets is the repeated, empirical encounter with patterns of objects in experience, and from these naive, intuitive representations (ultimately explainable by neuronal arrangements in our brains) we are able to proceed, theoretically, to create the entire conceptual edifice of mathematics. Maddy presents her reconstruction of mathematical realism as compatible with a physicalist ontology – since her sets of medium-sized objects are indeed spatio-temporally located entities – and, more radically, argues that it implicitly imposes an ontological monism. She writes that

> [i]n place of the old picture – physical reality here and now, mathematical reality nowhere and nowhen – set theoretic monism offers a spatio-temporal reality inseparably physical and mathematical. Physics and mathematics, on this new picture, are two sciences, along with chemistry, biology, psychology, and the rest, that study aspects of this reality. Each science has its own vocabulary and laws, its own techniques and methods, but this doesn't mean that the world itself is divided into the physical, the mathematical, the chemical, the biological, the psychological, and so on. *Rather, everything is ultimately physico-mathematical or mathematico-physical.* (1990: 158; emphasis added)

This is a seductive thesis, one which seems to offer some starting point, for the reader of Badiou's philosophy, to make naturalist sense of his mathematical ontology. However, Maddy's position is still unsatisfactory: the empirical grounding of our intuitions about sets (and mathematics as a whole) in physical objects does not have much explanatory power when confronted with the Wignerian riddle of how mathematics can be applied to predict natural phenomena. For to stop at the identification of the source of mathematical thought in experience can only lead to a mathematical constructivism that, in my view, fails to deliver a strong explanation of how high-level mathematical formalisations can be unexpectedly applied to physical phenomena. This would be a slippery slope towards the conclusion that the mathematisation of natural sciences happens to work just because mathematics is the cognitive machinery which we have developed to make sense of the world. Moreover, it would forfeit any realist ambition, surrendering our metaphysics to a Brouwer-like intuitionism or a reductionist position wherein *all* mathematical objects are identified with physical states (neuronal activation patterns) of the brain.

If Maddy's claim that everything is ultimately physico-mathematical is limited to the observation that our mathematics is drawn from our encounter with the physical world (and not from a realm of abstract eternal objects) – a claim I endorse – Philip Kitcher (1983) offers a critique of mathematical a priorism and Platonism that introduces a reconstruction of the *historical* emergence of mathematical knowledge. Kitcher argues that mathematical knowledge originates from ordinary everyday perception and that more sophisticated forms of mathematics have been historically developed, following a broadly Kuhnian pattern of discontinuous evolution, through the rational modification of these experience-based mathematical practices. Like Maddy, he holds that our first encounter with mathematical objects depends on an encounter with physical realities, so that, for example, 'children come to learn the meanings of "set", "number", "addition" and to accept basic truths of arithmetic by engaging in activities of collecting and segregating' (1983: 107). Kitcher writes that 'as a first approximation, we might think of my proposal as a peculiar form of constructivism' since

> [l]ike the constructivists I hold that arithmetical truths owe their truth (at one level) to the operations we perform ... Unlike most constructivists, I do not think of the relevant operations as private transactions in some inner medium. Instead, I take as paradigms of constructive activity those familiar manipulations of physical objects in which we engage from childhood on. Or, to present my thesis in a way which will bring out its realist character, we might consider arithmetic to be true in virtue not of what we can do to the world but rather of what the world will let us do to it. (1983: 108)

Rather than grounding his thesis on the contingent abilities of particular human beings, Kitcher envisions mathematics as the 'constructive output of an ideal agent' (1983: 109) endowed with normal cognitive abilities but unlimited resources of time and space. The higher concepts of abstract mathematics are then arrived at via idealisation from the empirically based capacities of this ideal agent. Kitcher then integrates this thesis with an historical dimension, illustrating the process of mathematical theory-change with a case study of the evolution of calculus between the seventeenth and the nineteenth centuries. However, as in Maddy's case, simply holding that, in some historically complex way, mathematics generalises from our experiences of the external world does not offer an answer to the problem of the predictive power of mathematics. It still fails, that is, to offer an explanation of how newly discovered, non-directly observable entities happen to be explained and predicted by mathematical formalism. We need an ontologically more daring stance in order to resolve this problem, even when endorsing Maddy's and Kitcher's empiricist spirit. As J.S. Mill (whose own empiricist view of mathematics as developed from enumerative induction is, Frege notwithstanding, an eminent forebear to Maddy's and Kitcher's stances) so eloquently wrote,

> [t]he notion that truths external to the mind may be known by intuition or consciousness, independently of observation and experience, is, I am persuaded... the great intellectual support of false doctrines and bad institutions. By the aid of this theory, every inveterate belief and every intense feeling, of which the origin is not remembered, is enabled to dispense with the obligation of justifying itself by reason, and is erected into its own all-sufficient voucher and justification. There never was such an instrument devised for consecrating all deep-seated prejudices. (1981: 233)

An interesting dialectical confrontation of mathematical Platonism and neurobiological empiricism is performed by (respectively) Alain Connes and Jean-Pierre Changeux (Changeux and Connes 1995). The former, a mathematician, defends his belief in an '*archaic mathematical reality*' (1995: 182), a realm of mathematical objectivities independent from both our minds and external physical reality, so that the truth of our mathematical statements is constrained by the objective states of affairs in this mathematical realm (mathematical facts are *discovered*, not invented). The latter, a neuroscientist, wants to deny the pre-existence of a mathematical reality to our experience and to reduce mathematical objects to 'dynamic physical states of neuronal networks' (1995: 183), thus stripping them of any abstract character and explaining our employment of mathematical methods to describe reality as an outcome of Darwinian evolution. Evolutionary processes endowed humans with areas of the brain responsible of a range of particular functions, from the basic assessment of numerical quantities – an ability we share with certain animals[34] – to the construction of complex mathematical objects through specialised neural networks. In the course of this conversation, Connes comes around to agreeing with Changeux that indeed this neurobiological account can offer insights into the workings of the mathematician's brain, but insists that the coherence of the productions of neuronal assemblies depends on the extent that they 'match up with the structure of external mathematical reality' (1995: 116): in his view Changeux is guilty of 'a confusion between conceptual tools and [mathematical] reality' (1995: 30). Against this stance, Changeux reiterates that the 'coherence is internal to both the brain and mathematics because mathematics exists *in* the brain of the mathematician' (1995: 125): there is no metaphysical gap between the operations of the brain and the objects it studies.

Both these positions have a certain explanatory purchase on the 'phenomenon' of mathematics, but in both insurmountable shortcomings surface. Connes' Platonism is, as he is happy to admit himself, a matter of *belief* in an unobservable (at least without a special, *ad hoc* faculty of intellectual intuition) mathematical realm. This leads him to the brink of a repugnant mathematical idealism when claiming that 'prime numbers ... as far as I am concerned constitute a more stable reality than the material reality that surrounds us' (1995: 12). On the other hand, the mind-independence of the mathematical realm remains the best explanation for the 'discovery' of new mathematical truths. Certain properties of mathematical objects (from the peculiar characteristics of prime numbers to those of the finite simple groups referred to by Connes (1995: 19)) seem to defy a neurobiological explanation: what would the evolutionary advantage of being able to establish the twin primes of perfect numbers be? Even when acknowledging the generative character of mathematics, Changeux links it to the more general 'cerebral faculties connected with the use of human language'[35] (1995:19) and ultimately defends the constructivist thesis that the mathematician 'is constructing a universal language that permits him to recognize properties of the object he constructed in the first place ... he "discovers" only the consequences of what he himself has conjured up!' (1995: 20). The *rationale* of this stance, however, is correct. Changeux, like myself (and Badiou), is an immanentist and a materialist who wants to reject any supernaturalist reference to 'what might be called the "mythology of origins"' (1995: 179) and 'mythic residues of Platonism' (1995: 25) underlying the claim for the priority of a diaphanous mathematical realm. Moreover, he correctly notes that an appreciation of the *historical* (contingent) evolution of mathematics allows us to do away with finalism (1995: 36) and to 'secularize' mathematical objects (1995: 18).

(Note, however, that one point on which both Connes and Changeux agree is that natural – external – objects and mathematical objects cannot be identified. Their respective reasons, however, could not be more different. Changeux's rejection of this thesis is no surprise, since for him, as mentioned above, mathematical object are internal physical objects, that is specific, material cerebral states. Connes, on the other hand, is keen to preserve a strict separation between the physical and the mathematical. He argues that 'external physical reality is a part of archaic mathematical reality', that indeed this mathematical reality 'exists on the same footing as external physical reality' and that 'it's the physical universe that is inside archaic mathematical reality' (1995: 206). Connes ends up defending, as Changeux correctly notes, 'a sort of dualism...between matter and mathematics that reminds me of the Cartesian distinction between mind and body' (1995: 55).)[36]

Changeux's neurobiological account of the genesis of mathematics should be integrated with an assessment of the central role played by cognitive technologics (external artifacts employed by humans to facilitate computation),[37] first of all *writing*. Historian of mathematics Reviel Netz has long stressed the importance of pursuing a project of 'cognitive history' (in the specific context of Greek mathematics but this project, *mutatis mutandis*, could be generalised) defined as the 'study of culturally specific practices, in which universal human cognitive abilities are assembled together and implemented with the aid of specific tools and technologies' (2002: 321). He has turned his attention specifically to the ways in which diagrams and texts (and by extension the *material* substrate on which diagrams were drawn) were an indispensable and mutually supposing part of Greek mathematical geometry. Lettered diagrams would facilitate communication between the community of mathematicians and indeed enjoy logical priority over written or oral exposition, to the point that they would be a necessary and irreplaceable part of the understanding of a geometrical proof (see Netz 1999: Chapter 1).[38] More generally, and reaching further back in human history, there is ample evidence that basic quantificational inscription (on material substrates) like tallying on counting sticks developed contemporaneously with (or arguably *preceded*) the ability to articulate number concepts verbally.[39]

Somewhat more speculatively, it seems to me a reasonable hypothesis that some such kind of inscriptional tallying (on rocks, bones, wood or whatever other material medium) could easily have been the first kind of *writing* that the human species ever developed (to leave a mark for each day until the next full moon, for example, seems a much more basic priority for a prehistoric human than the relatively leisurely activity of cave painting). Even recognising the difference in scope between a tallying stick and a proper calculation tool (like an abacus), it seems fair to say that *number*, more than anything else, creates the need for an external, material support (because of the limitations of the human brain to manipulate large quantities).[40]An extensive account of the biological evolution of our mathematical capabilities has been offered by Stanislas Dehaene (a colleague of Changeux – their Dehaene and Changeux 1993 being a landmark publication for the neurological study of numerical abilities) in his *The Number Sense* (Dehaene 2011). Dehaene's fundamental Darwinian thesis – which, against Badiou, I endorse as a naturalistic explanation of mathematical *thought* – is that

> we are...endowed with a mental representation of quantities very similar to the one that can be found in rats, pigeons, or monkeys. Like them, we are able to rapidly enumerate collections of visual or auditory objects, to add them, and to compare their numerosities. I speculate that these abilities not only enable us to quickly work out the numerosity of sets, but also underlie our comprehension of symbolic numerals such as Arabic digits. In essence, the number sense that we inherit from our evolutionary history plays the role of a germ favoring the emergence of more advanced mathematical abilities. (2011: 28–9)

Dehaene's book was originally published in 1997, seven years after Maddy's tentative (and mostly speculative) Hebb-inspired suggestion that we should look into the neurophysiology of the brain to explain how we come to develop the fundamental intuitions (and then the axioms) of set theory. Dehaene could therefore ground his hypothesis on a number of experimental results that were simply unavailable in the late 1980s. Today, as Dehaene points out in the postface to the 2011 edition of his book, even newer technologies unavailable in the 1990s (mainly functional magnetic resonance imaging (fMRI)) have allowed for the discovery of 'single neurons that code for number in the monkey brain, at a precise site in the parietal lobe that appears to be a plausible homolog of the human area that activates when we calculate' (2011: x), an area called, in the human brain, the 'interparietal sulcus' (2011: 239).

Expanding on Dehaene's work, Susan Carey has defended the existence of two rather different systems of core cognition with numerical content: analogue magnitude representations (based, roughly, on the approximation of magnitude) and parallel individuation (based on individuals) (2009: Chapter 4). Operational differences notwithstanding, both systems depend on representations of sets (taking sets of individuals as input) and support computations over sets, and in both cases '[t]he domain-specific perceptual analyzers that encode number, as well as the arithmetic computations defined over the resulting representations, are evolutionarily ancient, most likely innate, and operate throughout the life span' (2009: 134). Moreover, Carey argues for (and offers a model of) a discontinuity in the capabilities of our conceptual systems (including, but not limited to, numbers): from the innate systems available already to pre-linguistic infants, new systems emerge, affording resources that transcend those of previous systems. For example, Carey highlights the progress from the basic representation of cardinal values of sets to a three-year-old ability of representing positive integers, and further argues that 'a numerical system that encompasses fractions and decimals is incommensurate with the preschool child's hard-won numeral list representation of the positive integers' (2009: 413).

The lesson to be learned from this survey of work in the philosophy and cognitive science of mathematics is that the cognitive neurosciences are today fully equipped with the conceptual and technological resources to pursue research project on the brain-dependent conditions of possibility of our mathematical cognition. The results of this project will become more and more accurate as the imaging power of our technological apparatuses improve, and no philosopher can be excused for ignoring them. Devotion to Plato should give way to respect for the empirical sciences, but the disclosures of the latter on the neurological and socio-historical genesis of mathematics should be reconciled with a realist explanation of the applicability of mathematics to the physical world.

2.5 Two naturalising strategies

I should now begin to flesh out my own position. If mathematical realism amounts to a commitment to an abstract, non-spatio-temporal and causally disconnected realm of mathematical objectivities (mathematical

Platonism), I am not a realist. But a sweeping neurobiological reductionism won't do either. A robust philosophy of mathematics needs to offer a coherent explanation for the effectiveness of mathematics in the natural sciences without falling into anthropocentric (intuitionist, constructivist or crypto-theological) traps. Mathematical realities (as opposed to concepts), then, should be seen as enjoying the very same perceptual and conceptual independence as physical reality (assuming, of course, a metaphysically permissive scientific realism). The proposal is to conceive of the two as really *one and the same reality*: there is no qualitative distinction between the discovery of unobserved regions of the universe or of unknown subatomic particles and the discovery of hitherto unknown mathematical structures, for these are only distinct strata or patterns of the same reality.

I take this to be a materialist, immanentist and naturalist position. It is naturalist because it refuses to refer to any extra-mathematical cause and because it envisions mathematics, physics and philosophy as being on an explanatory continuum (as I will explain below, one that, however, does not enforce a strict reductionism). It is immanentist because it refuses any kind of transcendence for mathematical entities and collapses the physical and mathematical to the same immanent plane. That of materialism is probably the most controversial characterisation of this position, since it apparently undermines any intuitive understanding of 'matter' to a particular instantiation of a mathematical structure. I will henceforth employ the somewhat audacious neologism 'matherialism' to express the ontological monism (one world-ism) supported by this hypothesis. With Badiou, I agree that the standard definition of mathematical Platonism leads to meaningless debates regarding the existence of a world of mathematical idealities, and that mathematics offers the grammar for ontology. However, against Badiou, I also am ready to claim that the structures presented in mathematical thought are all that there ultimately is to everyday, material reality, without gaps between Being and World.

The apparently irreconcilable contrast between idealising Platonism and reductionist neurobiological explanations can thus be resolved. A middle way, rejecting the existence of a 'noetic world of supernatural entities' (Polkinghorne 2011: 34) while granting a mind-independent existence to the object of study of the mathematician, allows for a deflationary resolution of the problem of the applicability of mathematics to the physical world (and, once cast in this light, the Badiouian problem of the relationship between the ontological and the empirical) and forecloses supernaturalist responses deferring such an explanation to the providential schemes of a divine subject.[41] Like Badiou, I want to endorse a realist account of mathematics capable of explaining its truth-producing power while refuting the naive 'Platonist' belief in the existence of mathematical objects. At the same time, unlike Badiou, I want to reject a rationalist insistence on mathematics as a pure form of thought and an anti-naturalist, elitist disdain for biological and historical[42] explanations. To correct Badiou's noocentrism with, on the one hand, a naturalised mathematical epistemology and, on the other, a structuralist philosophy of mathematics amounts to a *vindication* of Badiou's ambitious understanding of mathematics as foundational intellectual practice. My aim is to preserve mathematics as the highest form of thought, but through a naturalist demystification of its origin, placing it on an immanent continuum with the rest of reality. In other words, there are two directions from which to recuse the distinction between thought and object: Badiou's rationalism does so by collapsing mathematical thought into its scriptural production, my empiricism does the same by generalising this conflation on a cosmic scale.

The project of offering an explanation for the applicability of mathematics to the empirical world does *not* coincide with an account of how human beings are able to acquire the ability to manipulate numbers and shapes. So, two theses can be simultaneously endorsed. On the one hand, the entirety of our mathematical *concepts* arises from the empirical encounter with the physical world and is propelled (on a biological level) by Darwinian evolution and the neuroplasticity of our brains, and (on a cultural level) by the socio-historical transmission of mathematical techniques. On the other, the best explanation for mathematics' success in describing the unobservable world should abductively lead us to infer that, at irreducible scales, there is nothing more to ('physical') reality than the mathematical structure used to describe it, and all such structures are ultimately depth-less iterations of the void.[43] The two strategies, the biological, bottom-up explanation and the structuralist, top-down abduction, are not mutually incompatible (or at least not evidently so, an argument needs to be given for their incompatibility). A bottom-up neurobiological naturalisation of mathematics is compatible with a top-down insight into the mathematical nature of physical reality once we recognise that, while conceptual thought is *logically* prior in our exploration of the universe, Being *qua* objective natural reality is *ontologically* prior as the condition of possibility for the emergence of thought. Having sketched the possibility of the former strategy, in Chapter 4 I will turn my attention on the latter, proposing a coaptation of mathematical and physical structuralism in order to show

how a structural realist re-reading of Badiou's ontology can solve the problem of the ontico-ontological relation by postulating a structural/mathematical reality.

2.6 The matherialist option

I opened this chapter by expressing my reservations concerning a pressing problem in Badiou's system, one caused by his uncompromising rationalism: the unaccounted for ontological-ontic relationship. The lack of an adequate explanation for how a mathematical ontology can bear on empirical, concrete situations, and the reference to an 'inconsistent multiplicity', undermines his immanentist ambitions. Even if we follow Badiou's recommendations and avoid misinterpreting his inconsistent multiplicity as a Deleuzian virtual reservoir – which, for Badiou, is but another avatar of the One and thus 'a secularized, or sublimated, God, over-existing puppeteer of being' (2002: 66) – it is easy to remain unsatisfied with the emptied yet preserved category of pre-presentational inconsistency. Badiou's demotion of the realm of the sensible in favour of the abstract realm of mathematical intelligibility makes it problematic to square his project with the empirical practice of contemporary science.[44] On the one hand, I have proposed to supplement Badiou's shortcomings through a foray into relevant trends in the philosophy and science of mathematics, offering a demystification of our mathematical abilities. On the other hand, Badiou's identification of mathematics with ontology can be naturalised by collapsing the quasi-Heideggerian[45] dichotomy of presented immanence and inconsistent Being: the hypothesis, to be further substantiated in Chapter 4, is that there is nothing more to what there is than what is presented, and this is ultimately a mathematically describable structure.[46]

To sum up, and chart the conceptual terrain, the relationship between mathematics and the physical world can be conceived in four ways:

1. Thought and Being are really one and the same; mathematics, being the highest form of rational thought, articulates, without remainder, all that can be said about Being. Physics, on the other hand, is a science which regards the experienced world of phenomena (a particular domain of presentation) and its mathematisation is nothing but a formal exigency which does not imply any congruence between its object, 'matter', and that of mathematics, 'Being'. This is Badiou's approach (and, in his reading, Plato's).

2. Mathematics is an endogenous evolutionary product of the human brain, originating from the encounter with the empirical world and developed thanks to the exogenous effect of cognitive technologies such as counting tools and writing. Its applicability to the natural world is no mystery given that our cognitive machinery processes inherently non-mathematical natural phenomena as mathematically describable. This is the stance defended by Changeux, Dehane and Johnston (and, up to a point, by Maddy) in terms of neuronal functioning of the brain and by Kitcher, Netz and Carey in terms of the historical evolution of our cognitive processes. All of these stances agree that mathematics is an intellectual production based on the experiential encounter with macroscopic physical objects in our living practices, what Lakoff and Núñez (2000) defined 'embodied mathematics'. Appropriately relaxing the notion of 'subject' so to include both the neurobiological subject *and* phenomenological transcendental subjectivity, to this group could be merged those phenomenology-based stances of Husserlian inspiration, holding that the objectivity of mathematical objects is transcendentally constituted by our own world-engaged cognitive-gestural acts.[47]
3. There exists a (non-mental *and* non-physical) realm of entities which is the domain of mathematics. Some of these objects will find application in theories about the physical world whilst others will remain part of the furniture of the mathematical realm, accessible only through some form of intellectual intuition. This is the position defended by Connes, Penrose and Gödel.[48]
4. The physical universe is (at least on some scale) amenable to mathematical description because it is (in some sense) inherently mathematical. No fundamental distinction obtains between physics and mathematics.[49]

I have intentionally left this last thesis (the matherialist or 'Pythagorean' option) vaguely formulated here, for in the nuances of its terms reside fundamental metaphysical differences. In particular, it should be noted that such a thesis is often hijacked by those interested in explaining the 'miraculousness' of the applicability of mathematics through divine intervention since, so the argument goes, the rational organisation of the universe can only be fully explained by the intervention of a rational organising agent concerned with our cognitive limits. This is the theistic and anthropocentric stance defended by Steiner, Polkinghorne and Ratzinger. Needless to say, I wholeheartedly reject this position as rebarbative to reason, invoking a supernatural, transcendent *deus ex machina* responsible for a human-friendly rational organisation of the universe.[50] The challenge for my own matherialist position, then, is that of defending the claim that at an elementary level there is no difference between mathematical and physical structures *without for this reason invoking some primordial intelligibility, rational organisation or pre-established harmony of the universe* (thus avoiding the pitfalls of either theism or idealism). This is, of course, a delicate position to defend since, *prima facie*, a human-independent mathematically describable structure seems to imply (or at least leave open the possibility for) the kind of rational organisation that theists want to detect in the universe as a proof for its divine origin: a *mathesis universalis* made possible by God the geometer.

Yet, that is a wager of this book: the possibility of disentangling a structural realist understanding of physical and mathematical reality (one and the same) from theological (or onto-theological) and idealist (transcendental or Hegelian) presuppositions or reappropriations, and indeed to claim with Badiou that its imperviousness to *meaning* makes the mathematical intrinsically unyielding to theological appropriations. Badiou's most 'secularising' drive is the subtractive insistence on the lack of any substrate to presented Being, the lack of any meaning to material reality, and the lack of any transcendent origin for the discovery of truth (unfulfilled promise, as I'll explain below). My re-reading of Badiou opens the path for a realism that is *neither* a kind of passive, receptive expectation[51] of the encounter with a reality that 'puts us in the accusative' (to use a Levinasian turn of phrase) *nor* an active act of wholesale constitution of the reality that we experience. Once again, my favouring of mathematical structures as fundamental constituents of reality does not depend on a neo-positivist preference for formal languages *nor* on a Badiouian rationalist insistence on the purely intelligible character of mathematical formalisation. Naturalistic immanence is what, in my view, supports the matherialist option: a stance that carries with it both deference to our best epistemic enterprise (science) and the imperative of responsibility which derives from our being part of a purely immanent reality. As such, it is no idle speculation, but a thesis with definite consequences regarding our employment of reason and our place in the universe.

In the next chapter, I will elucidate further precisely the contested notions of realism and naturalism, in order to set the stage for the matherialist thesis exposed in Chapter 4.

3
Taking a Stance on Realism and Naturalism

In the previous chapter, I identified a serious problem with Badiou's ontology in its failure to properly account for the difference between ontological and non-ontological (empirical) situations. I attempted to correct this deficiency through a foray into the philosophy and cognitive science of mathematics, with the intent of naturalising Badiou's understanding and employment of mathematics. In order to flesh out in more detail what this naturalisation entails it is time to shift gears and turn, in this chapter and the next, to a problem that I have left, so far, under-examined: realism.

Realism is one of those long-standing philosophical issues that have traversed the history of philosophy inflaming the intellectual fervour of its most gifted thinkers, who have variously attempted to articulate the problematic relationship between *appearance* (to us) and *reality* (in itself) – either arguing for the inescapable horizon constituted by the former (anti-realists) or seeking to obtain access to the latter (realists).[1] It sometimes seems that little resolution has been achieved in the last millennium or two since it is easy to find, in recent publications, debates that would have been familiar (terminological nuances notwithstanding) to a first-century CE Greek or Indian philosopher. The longevity of this problem is testimony to its central importance in our self-understanding as cognitively active beings, located in a universe about which we aspire to achieve knowledge. My intention, as intimated in the Introduction, is to draw upon resources from both analytic and continental sources in order to draw a coherent, non-theological metaphysical picture capable of supporting a robust *scientific realism* (a realism with regard to the domain of things studied by the empirical sciences). Different realisms are defined according to the domains of objects the reality of which they predicate: propositions, moral values, meanings, modality, numbers, and so on. I will be concerned with the objects of science and mathematics, since my matherialist/structuralist position collapses, at irreducible scales, the domain of physical and mathematical object together.

3.1 Scientific realism

There are as many definitions of scientific realism as there are volumes written on the subject, where tentative definitions are often broken down into separate core commitments. Perhaps the most comprehensive way to enumerate them is to employ Stathis Psillos's three-tiered definition, breaking scientific realism down to core theses or stances which realists should be committed to:

1. The metaphysical stance asserts that the world has a definite and mind-independent natural-kind structure.
2. The semantic stance takes scientific theories at face-value, seeing them as truth-conditioned descriptions of their intended domain, both observable and unobservable. Hence, they are capable of being true or false. Theoretical assertions are not reducible to claims about the behaviour of observables, nor are they merely instrumental devices for establishing connections between observables. The theoretical terms featuring in theories have putative factual reference. So, if scientific theories are true, the unobservable entities they posit populate the world.
3. The epistemic stance regards mature and predictively successful scientific theories as well-confirmed and approximately true of the world. So, the entities posited by them, or, at any rate, entities very similar to those posited, do inhabit the world. (1999: xvii)

In this chapter and the next I will focus mostly on the metaphysical and epistemic aspects (reserving a more focused concern with semantic considerations for Chapter 5).[2]

The scientific realist is committed to the objective and mind-independent existence of those entities that the best (methodologically regulated)[3] epistemic practices refer to in their most experimentally confirmed and predictively successful theories. Mind-independence can be cashed out both in terms of the independence of the perceived from the perceiver (against Berkeleyan idealism) and in terms of the independence of the conceptualised from the concept-monger (against Platonic or Hegelian idealisms). Objective existence, on the other hand, indicates how the primary qualities[4] of the entity in question will be observed to be precisely the same by any inquirer at any time (provided that such properties do not *independently* change with time: objectivism does not entail eternalism). Both mind-independence and objective existence are *conjointly* necessary clauses, in order to save the reality of the phenomenon from falling into the conceptual cracks between subjective and objective idealism.

At its most basic, scientific realism entails an ontological commitment to the existence of *unobservable* entities.[5] Even the special sciences, the objects of which are in the observable realm, the realist holds, obtain their explanatory power (at the end of the reductivist chain) thanks to a reference to causal mechanisms operating at the unobservable scale. As such, scientific realism is an *extension* of common-sense realism about observable items of everyday experience through experimental apparatuses, which supplement and improve upon sensory capacities.

To do some conceptual house-keeping let me taxonomise realist stances thus: (1) *metaphysical realism* is the broadest taxon, describing the stance of commitment to mind-independent entities (of whatever nature); (2) *common-sense realism* describes the stance of commitment to the mind-independent existence of all those entities which our unimpeded perceptual capacities allow us to have contact with in our everyday dealings with the world; (3) *scientific realism*, finally, is a more refined and specialised thesis about (an idealised model of) scientific practice, entailing the ontological commitment to all those entities that can be the objects of scientific inquiry.

The Sellarsian metaphor of a 'stereoscopic vision' (1963: 4) sought by philosophy – the indispensable task of constructing an epistemic interface between the two conceptual systems of manifest and scientific image[6] – here supports the idea that the scientific and the everyday perception of reality, in all their qualitative differences, refer to one and the same reality. The main challenge for this view, of course, is that of specifying the relative ontological status of the observable and the unobservable world or, again in Sellarsian terms, that of allowing for the (ontological) reducibility of the manifest image to the reality described by the scientific image while preserving the relative (conceptual) independence of our everyday dealings with the human-scale world (including the normative sphere).

In other words, the scientific realist must be committed to the idea that scientific knowledge can be placed on a continuum with common-sense knowledge from a perceptual standpoint (so that it can be meaningful to say that the physicist non-inferentially *perceives* a particle through the mediation of experimental apparatus in the same way she non-inferentially *perceives* the pen used to take her notes) while at the same time constituting a radical break with it on the conceptual level (the everyday experience of time-enduring, persistent, solid objects being in irreconcilable conflict with the status of these very same objects from the micro-physical point of view – Eddington's two-tables dilemma).[7] Yet the scientific reductionist cannot be a realist in the lab and a sceptic in her backyard, but needs to conceive of her two activities (say, smashing particles and cutting wood) as interactions with different aspects or levels of the same reality. It should be said that *from within the manifest image*, the manifest and the scientific image are irreconcilable while *from within the scientific image* they are perfectly continuous (see Sellars 1963: 6), once manifest observables are reconceived as (and eventually even directly *seen as*) composed of unobservables particles. The reconciliation of these two standpoints is what Sellars referred to as (the *regulative ideal* of) a stereoscopic vision of reality, recognising the ontological and explanatory priority of the scientific image while admitting the phenomenologically and normatively irreducible character of manifest image practices.

Scientific realism is as much a thesis about the universe studied by science (mind-independent entities), as it is a thesis about science itself (a truth-seeking methodology of empirical inquiry), and indeed one about human inquirers. Regarding the latter, scientific realism is an anti-anthropocentric position to the extent that it reserves no special status (either positive or negative) to the epistemic abilities of human observers:[8] ontological inquiry into reality is constrained by epistemological circumspection, but cannot be regulated by human perceptual limitations. Epistemology does not dictate ontology.

The scientific realist *does* recognise that scientific inquiry is conditioned by certain *cognitive* structures and evaluative standards. Knowing is not limited by perception for the simple reason that it is through conceptual mediation that we truly acquire knowledge. Against all metaphysical (onto-theological) notions of an intrinsically rational *ordo naturalis* the realist rejects the necessary match between concepts and the objects in the universe (the *givenness* of the categorical structure of reality). Against the rationalist of either Cartesian, Leibnizian or Hegelian stripes, the realist places limits on the power of empirically unconstrained pure reason. Unlike Kant, however, she argues that these limits are not a-temporal and invariant transcendental conditions,[9] but – adopting a relativised notion of *a priori* principles – are object of historical progression and (on a larger time-scale) are determined by our evolutionarily developed cognitive make-up. Naturalistic monism is not incompatible with a methodological dualism between concepts and objects. This can be defined as a 'transcendental realism' to the extent that we recognise the logical and temporal priority of the real to that of the conceptual: the cognitive capacities of human beings are part of and dependent on the prior reality of the universe, and can be employed to discover objective features of it.

We are creatures evolved to achieve reproductive fitness in (and navigate our way around) a world of middle-sized objects[10] and our basic conceptual toolbox developed to facilitate this goal before that of uncovering the deep structure of reality. Through the historical development of rational thought we have pushed the horizon of ignorance back a long way through the development of methodologically regulated empirical investigation, the results of which have fed back into (and improved) our conceptual resources: an impressive achievement indeed. But it would be rationalistic hubris to assume that we have already uncovered all that there is to know.[11] If our present cognitive limitations do not curtail *a priori* our ambitions to obtain an accurate match between thought and Being, the responsible stance for the naturalistically inclined realist is to suspend judgement about the quasi-eschatological fulfilment of this fully rational deployment and rather aim at a piecemeal and asymptotic pursuit of an increasingly approximate truth via a continuous empirical revision of conceptual frameworks. We can't presently *know* how we are going to improve our ways of *knowing*.

This position does not open a backdoor for supernaturalism: that some facts about the universe are as yet unknown or even unknowable does not entail a rejection of an evidence-based scientific methodology. The methods of hypothesis-formation and empirical experimentation on which science is based are demonstrably the best methods we have to achieve knowledge: 'in the dimension of describing and explaining the world, science is the measure of all things, of what is that it is, and of what is not that it is not' (Sellars 1963: 173). But to grant science this role does not, once again, imply that its inquiry is a process with a definite end: the boundaries of the unknown might be forever pushed back and yet, like an epistemic horizon, forever remain out of reach.[12]

3.2 The realist stance

Borrowing a term made famous, in this context, by Bas van Fraassen, I take realism to be better described as a *stance*. Van Fraassen describes the difference between his own stance and that of the materialist as that between the epistemically humble attitude of the (constructive) empiricist and the dogmatic and overweening metaphysical realist. The latter position (in Putnam's famous three-pronged definition) is that according to which 'the world consists of some fixed totality of mind-independent objects. There is exactly one true and complete description of "the way the world is". Truth involves some sort of correspondence relation between words or thought-signs and external things and sets of things' (Putnam 1981: 49). I have a very different understanding of both realism and of metaphysics to those of van Fraassen and Putnam,[13] but I am happy to adopt the term 'stance' insofar as it is meant to index a philosophical position that 'may involve or presuppose some beliefs…but cannot be simply equated with having beliefs or making assertions about what there is' (van Fraassen 2002: 48).

Van Fraassen takes metaphysics as an illegitimate extension of empirical science, an attempt to 'continue science so as to answer questions scientists do not ask but in the same way as science answers its questions' (2002: 11) which inevitably ends up in 'a tissue of falsehoods' (2002: 16) in the name of explanation. He charges realist metaphysics with a naïve, if not downright childish desire for 'deep' explanations beyond pragmatic ones: explanation, the empiricist claims, is 'less a virtue than an anthropocentric pleasure' (van Fraassen in Rosenberg and Balashov 2002: 69).[14]

Van Fraassen's realists are thus busy irresponsibly postulating unknowable phenomena beyond empirical appearances to quench their explanatory thirst, arranging these metaphysical entities in order to offer 'something that purports to be the "One True Story of the World"' (in Hilgevoord 1994: 116). To employ Herbert Feigl's vivid turn of phrase, the empiricist shuns 'the sham completeness metaphysicians procure for their world pictures by verbal magic' (in Feigl and Brodbeck 1953: 13). Against such metaphysical extravagance, then, the empiricist adopts a *stance*. Differently from the metaphysician, the empiricist does not harbour systematising ambitions nor any specific belief or *a priori* conviction regarding the nature of reality – the term stance (as opposed to *thesis* or *belief*) is supposed to express this open-minded, revisable and noncommittal attitude. Interestingly, van Fraassen singles out materialism as a 'stance misunderstood': surviving under the 'false consciousness' of offering factual claims regarding what there is (something which it is actually unable to do), metaphysics is actually best identified as another stance, characterised by van Fraassen as defending a dogmatic (and vacuous) belief in the truth and completeness of science's current best theories. While allowing that empiricism shares with materialism an 'admiring attitude' towards science, van Fraassen stresses that this attitude is 'not directed so much to the content of the sciences as to their forms and practices of inquiry' (2002: 63). Summing up his argument then, the claim is that

> [f]or the materialist, science is what teaches us what to believe. For the empiricist, science is more nearly what teaches us how to give up our beliefs. All our factual beliefs are to be given over as hostages to fortune, to the fortunes of future empirical evidence, and given up when they fail, without succumbing to despair, cynicism, or debilitating relativism. (2002: 63)

There is nothing I find disagreeable in this formulation of empiricism. In fact, I take it to be precisely the same set of meta-theoretical commitments for which the *realist* pursues her theoretical enterprise. I have already characterised the realist attitude as a radically immanentist one. If rejecting transcendence means rejecting any kind of intellectually reconstructed cause beyond all possible experience (perhaps only the object of a dubious intellectual *intuition*), the line separating my position and van Fraassen's regarding a commitment to unobservable entities seems to be a thin one indeed, especially when both stances are concerned with the role of explanation. The immanentist realist categorically rejects any reference to transcendent causes to explain phenomena: but does the rejection of transcendence entail the kind of anti-metaphysical attitude recommended by van Fraassen? My argument is that it does not, which is why I am content to characterise realism as a stance: realism cannot be reduced to a propositionally articulated belief that *p*.

Realism, certainly, is motivated at its core by an explanatory vocation: the best hope for coherent causal explanations of phenomena requires theories to refer to unobservable entities whose microscopic behaviour is responsible for the macroscopic appearance. That of explanation, however, is a rational imperative[15] which does not force the realist to carry the heavy metaphysical burden van Fraassen deposits on her shoulders. Van Fraassen constructs the realist position as descending directly from Aristotelian teleology and medieval speculations on natural necessity, inheriting an obsession with 'why questions', beholden to a metaphysical worldview of natural order and pursuing the comforting yet vain quest for an 'illusory sense of ontological security' (in Hilgevoord 1994: 123). For him, the realist is prejudicially committed to an *a priori* demand for explanations that are (as the jaded empiricist knows) nowhere to be found, an ultimately dogmatic or irrational position. Not all explanations, however, need be so loaded. The kind of realism that I will defend in this book is constructed with the explicit aim of offering explanatory power without subscribing to the crypto-theological metaphysical background assumptions[16] which van Fraassen sees as necessarily operative in metaphysical realism.

It is possible to have a realist stance that does not 'reify content' (van Fraassen in Hilgevoord 1994: 132) nor prejudicially endorses natural necessities while, at the same time, preserving the truth-seeking and self-correcting character of science in the pursuit of a causal explanation of phenomena. The disagreement with the constructive empiricist pivots on the attribution of 'transcendence': while van Fraassen defends a principled agnosticism (and a *de facto* anti-realism) vis-à-vis experience-transcending entities, the realist is committed to the existence of objects which transcend the epistemic powers of the subject,[17] but does not accept anything which transcends the whole of reality (or, in Deleuzian terms, the 'plane of immanence') opening the space to vertical causation. Realism rejects the transcendent imposition of order by disentangling physical *causes* in the universe from (in Sellarsian terms) the logical space of *reasons* within which rational inferential patterns operate. That is to say, we can employ normative talk when discussing *our* methods of empirical inquiry without for that reason implying that *reality itself* (as a real space of physical, 'horizontal' causes) displays an externally imposed ('vertical') normative organisation.

Here I expose a key nerve of my thesis, since I take order and organisation as deceptively straightforward philosophical terms, fraught with metaphysical subtleties and theological niceties, all too often unreflectively employed in realist literature. This is where I move yet further from orthodox analytic scientific realism. In the following passage, Howard Sankey summarises a background assumption that is more or less tacitly shared by almost all scientific realists (and the rejection of which is almost always taken as an immediate admission of anti-realism):

> [a] mind-independent reality may be an amorphous, unordered world. Such a world is hardly a world worth fighting for. The world in which the realist should believe is not just a mind-independent reality, though it is at least that. It is a world with structure and order. (2008: 7)

When it comes to this emphasis on order, the only exception in the realist camp is that of the philosophers linked to the so-called Stanford

School of philosophy of science. Scholars like Ian Hacking, Nancy Cartwright, John Dupré, Peter Galison and Margaret Morrison are agreed in their opposition to the 'Unity of Science' ideal which, after its logical positivist heyday in the first half of the twentieth century and its historicist undermining by Thomas Kuhn in the 1960s, still survives in contemporary philosophy of science. Stanford School philosophers share the conviction that the best science would be better guided by a metaphysical vision of the universe as a 'dappled world' (Cartwright 1999), that 'although unification is an important part of the scientific process, an analysis of how it takes place reveals that it can in some instances have very few, if any, implications for a reductionist metaphysics and an ontological unity of nature' (Morrison 2000: 5). Indeed they think that the belief that such a reductionist account 'must exist in principle, or in the mind of God, is at best an inference from what we do know about the behaviour of systems simpler by many orders of magnitude, and hence an inference that goes beyond any decent empiricist strictures. It is, in short, a super-naturalist belief' (Dupré in De Caro and Macarthur 2004: 50). I will return to these observations in Chapter 5, since the claim that my matherialist/structuralist proclivities can be reconciled with this anti-reductionist stance is a central proposition of this book.

Consistently with the above, I want to stress how scientific realism, as I construe it, is less of a doctrine and more of a research programme precisely because it does not entail a set of fixed ontological commitments but, on the contrary, a stance of open receptivity to what new, unforeseen and constantly improvable information regarding reality the natural sciences might deliver. (It is here that I think an encounter of analytic and continental positions can furnish conceptual help: Badiou's concept of 'truth' will precisely be brought to bear upon the necessity of defining how the sciences can come to achieve points of rupture with current best knowledge and aspire to a truth that requires a radical revision of our conceptual arsenal in order to be appraised and pursued). So, against the scruples of radical empiricists and their admonitions on the naiveté of metaphysical realism such a hybrid approach to realism allows one *both* to hold fast to values of epistemic modesty *and* to have the ambition (or, in Wright's term, *presumptuousness* (1992: 1)) of being able to increase our knowledge of the universe in a fallible, laborious but progressive fashion. This also means that, while the realist acknowledges that ontological and epistemological registers are, in her discourse, always conjoined (we cannot strive for objectivity about reality without acknowledging the ways in which our perceptual and cognitive skills are always undergoing conceptual mediation) the logical priority goes to a mind-independent reality which must be (transcendentally) taken as condition of possibility for our access to it.

3.3 Transgressive naturalism

To clarify my position further, I will sketch Lee Braver's argument (2012) for the existence of a realist current in continental philosophy,[18] which he christens 'Transgressive Realism'. Braver traces back the origins of this stance, amounting to a middle way between (pre-Kantian) realism and (transcendental) anti-realism, to Kierkegaard's response to the post-Kantian Hegelian project, and sees it maturing through the continental tradition in (certain themes of) Heidegger's and, more fully, in Levinas's philosophy. Reality in itself, the transgressive realist argues, cannot be exiled to an epistemically inaccessible noumenal realm *nor* can human reason be considered as immanently deploying an intrinsically rational real. A compromise position between transcendental and absolute idealism, it amounts to an acknowledgement of ontological modesty, holding that reality cannot be fully grasped by the raw power of reason since it exceeds conceptual capture, together with the epistemically more ambitious belief that it is precisely by way of this transgression that reality forces a rearrangement of our concepts, capable of only asymptotically approaching truth. Reality is knowable *as* transgressing knowledge, is grasped insofar as it offers *obstacles*, resisting epistemic enquiries. Braver sums up the core intuition behind the transgressive realist's stance – a certain identification of reality with novelty – when he writes that

> the new, if it is to be truly new, must take us by surprise; it must fall upon us like inspiration rather than emerging from conscious planning since any planning would employ the very concepts and approach we are trying to get beyond. (2012: 277)

In this context, novelty should be understood not in its *additional* form, as a piecemeal increase in our total knowledge, but in its more radically *subversive* role vis-à-vis current knowledge, something heteronomously forcing a rearranging of concepts. Braver's transgressive realism, then, characterises these experiences of novelty as 'paradigmatic points of contact with a reality unformed by human concepts, when a true beyond touches us, sending shivers through our conceptual schemes, shaking us out of any complacent feeling-at-home' experiential points of contact that 'do not get squeezed into our mental structures but instead violate them, cracking and reshaping our categories' (2013: 12). 'This violation' Braver argues 'is the sign of their externality since everything we conceive remains the offspring of our concepts and so retains a family resemblance with them' (2013: 12). Following Braver's insight, my suggestion is that if new knowledge about the universe is the principal aim of the natural sciences, large sections of twentieth-century continental philosophy offer resources for philosophy of science precisely to the extent that a thorough theorisation of the problem of (ontological) novelty[19] – mostly under the rubric of the *event* – and an emphasis on conceptual plasticity and creativity have been central concerns of this tradition. If, as Braver correctly argues, our frail conceptual schemes are under constant assault from the science-mediated clash with a rebellious reality, we need all the philosophical ingenuity we can get in order to re-mould them. To *either* buttress them stubbornly and declare them unchanging *a priori* structures *or* to let them crumble while breathlessly standing in awe of the irruption of the unconceptualisable are equally undesirable avenues to, respectively, anti-realist criticism and a-rational mysticism. For those keen to bridge the analytic–continental gap, however, things are not as straightforward as this. As I established in the first two chapters (vis-à-vis Badiou), the main obstacle to overcome when attempting an encounter between continental and analytic approaches to science is that of negotiating a mutually acceptable form of naturalism (this is all the more evident when considering that, in one of the passages I quoted above, Braver referred specifically to Heidegger and his insistence on *Dasein*'s passive receptivity to an *Ereignis* of Being; ontology, the naturalist realist holds against Heidegger, is *not* phenomenology).[20] The challenge is that of exploiting the ontological speculations of continental thinkers without subscribing to the anti-naturalism which often is an integral part of them. Relating this point directly to Braver's 'transgressive realists' (from Kierkegaard to Levinas through Heidegger), the problem amounts to this: how to allow for experiential encounters with reality 'transgressing our anticipatory processes' (Braver 2013: 12) without equating conceptual transgression with the ingress of some transcendent Being into immanent reality. As I observed, the naturalist rejects any such transcendence, and the novelty of reality must thus be understood as being an immanent irruption of the hitherto unknown within the causal order of the universe.

A constructive development of the philosophy of science requires us to strike the right balance between naturalism and (rationalist) speculation. As Norris explains with admirable clarity:

> [o]ne can see how impoverished (or downright unworkable) is a rationalist outlook unanchored in the modes of our being in the natural world, but also how crucially defective is any naturalistic outlook that fails to recognize the full extent of the mind's creative powers, that is to say, its indispensably active role in the diverse procedures of conceptual exploration and inventive hypothesis-testing... [A] rationalism pressed to the limits (and beyond) of consistent application will always at some point encounter the need to make sense of its worldly or historical placement as well as acknowledging, at least implicitly, the intimate tie between reason and the various enabling as well as limiting conditions of our physical or natural embodiment. (2010: 6)

Naturalist rigour and speculative creativity should indeed be one tight philosophical package: for those with a passion for labels, I would call such a position 'transgressive naturalism'. The more effective way to fend off anti-naturalist attacks (and note that anti-naturalist positions are not limited to continental circles), then, is not a fist-banging naturalist dogmatism but a self-critical movement of historically and scientifically informed speculation at the limits of naturalism. The former unyielding and defensive approach seems to offer to anti-naturalists only a misplaced feeling of occupying an intellectual high ground, thus allowing them to present themselves as open-minded truth-seekers unafraid to question cherished beliefs and entrenched dogmas, an ironic overturning of reality.[21] Naturalism, in other words, cannot just be adopted as a scientific dogma. To some extent I agree with Timothy Williamson's provocative claim (perhaps more the letter than the spirit, given Williamson's favouring of 'armchair philosophy') when he writes that

> I do not call myself a naturalist because I do not want to be implicated in equivocal dogma. Dismissing an idea as 'inconsistent with naturalism' is little better than dismissing it as 'inconsistent with Christianity'... I sympathize with one motive behind naturalism – the aspiration to think in a scientific spirit – [but] [n]aturalism as dogma is one more enemy of the scientific spirit. (2011: n.p.)

Naturalism should indeed be interpreted as more than a craven genuflection towards the idol of science, and I take the point of Williamson's questioning, later in the same piece, the ability of real human beings (philosophers and scientists alike) to avoid mistaking their imperfect and revisable best current science for the *regulative ideal* of perfectly rational scientific inquiry. I take naturalism to commit us to more (or perhaps *less*) than a set of methodological dos and do nots. Naturalism, as I see it, is a philosophical or meta-scientific framework, pivoting on *a commitment to immanence*. This does not make a *dogma* out of naturalism[22] to the extent that meta-scientific frameworks *can* be rationally assessed,comparatively evaluated, and judged as progressing – through constant revision – towards an increasingly accurate conceptual grasp of reality: 'inconsistent with naturalism' *cannot* be equated – as a reason for disqualifying a hypothesis about reality – with 'inconsistent with Christianity' to the extent that immanentist/naturalist commitments *can* be rationally preferred over Christian ones (or whatever other set of theological, transcendent commitments).

I feel ambivalent towards the 'liberal naturalism' recently defended by some philosophers (see De Caro and Macarthur 2004 and 2010), since I am not sure that there is any philosophical advantage in the postulation of a domain of entities that are neither supernatural nor scientifically describable (part of the causal nexus of the universe). If all liberal naturalism is saying is that some entities are *not yet* under the purview of scientific explanation (and it is the task of philosophical creative thought to step beyond science in order to bring them back under scientific scrutiny) it is in agreement with my own transgressive naturalism. If it is saying, as it often seems to, that some entities are in principle refractory to *any* scientific recapture I struggle to see how this does not amount to supernaturalism, effectively breaking the immanentist principles of naturalism (*other*-natural is supernatural enough). Liberal naturalism is expected to help to solve the 'placement problem', the necessity of finding a role for non-scientific items like norms, *abstracta*, numbers and so on in the scientific image: leaving the vexed problem of norms aside, my own position, as anticipated in Chapter 2 and further expanded in Chapters 4 and Chapter 5, side-steps the problem of a supposedly 'third realm' or mathematical entities, by reinscribing them into the physical domain.

3.4 Historicist excursus: Collingwood and Bachelard

My realist metaphysics ventures to be both naturalist and historicist. For the latter *desideratum*, I take a stance regarding metaphysics influenced by R.G. Collingwood (a more interesting philosopher than he is often made out to be, summarily relegated to the dusty corner of the history of philosophy called 'British Idealism'[23]). Positioning himself against both the dominant positivist anti-metaphysical *doxa* and the 'continental' resurgence of fundamental ontology of his times, Collingwood reconceptualises metaphysics as neither to be rejected as a referent-less flight of theoretical fancy nor to be equated with the study of an alleged Being *qua* Being. Metaphysics, Collingwood argues, studies *presuppositions.* Presuppositions are those assumptions tacitly at work in the formulation of scientific questions, and are by him divided into *relative* and *absolute*. A relative presupposition '*stands relatively to one question as its presupposition and relatively to another question as its answer*' (1940: 29); that is to say, they are presuppositions that take a propositional form and that can be answers to precise questions (that the sun will rise and shine tomorrow is a presupposition for most of my daytime scientific activities, but this presupposition is – in line of principle – questionable, and can be verified by astronomical inquiry). Relative presuppositions, then, admit of the question 'Why?'. An absolute presupposition, on the other hand, '*is one which stands, relatively to all questions to which it is related, as a presupposition, never as an answer*' (1940: 31). They are not verifiable, do not take a propositional form, and neither truth nor falsity can be predicated on them. These are deep-seated assumptions made by the scientist and the lay person alike which are, *by definition*, not open to revision nor questioning, composing the never explicit bedrock of rational processes: '[t]o be propounded is not their business; their business is to be presupposed' (1940: 33). The task of the metaphysician, Collingwood argues, is that of analysis intended as the activity of 'disentangling and arranging questions' and of 'detecting presuppositions' (1940: 39, 40), telling relative and absolute ones apart, and making of the latter the main object of inquiry without assuming them to be susceptible of either truth of falsehood. Metaphysics, in sum,

> is the attempt to find out what absolute presuppositions have been made by this or that person or group of persons, on this or that occasion or group of occasions, in the course of this or that piece of thinking. (1940: 47)

As such, and this is Collingwood's central claim, '[a]ll metaphysical questions are historical questions, and all metaphysical propositions are historical propositions' (1940: 49), so it follows that 'the metaphysician is a special kind of historian' (1940: 62). Collingwood is well aware of the low esteem in which this historicist stance was held by his contemporaries,[24] and pointedly notes that '[i]n modern Europe absolute presuppositions are unfashionable. The smart thing to do is to deny their existence' since 'people are apt to be ticklish in their absolute presuppositions' and indeed 'when an absolute presupposition is touched, the invitation [to justify it] will be rejected, even with a certain degree of violence' (1940: 31, 42, 44). Collingwood, perhaps also playing on the ambiguity of the term *analyst* assigned to the metaphysician, speaks of this hostile reaction as a 'neurotic' habit (1940: 46) and a 'symptom' (1940: 44) that the presupposition touched is absolute and hence not open to revision. Note, however, that Collingwood here is merely vouching for the necessity of bringing absolute presuppositions to light through a painstaking process of self-examination of one's thought-patterns in order 'to have them scientifically described' and *not* examined, criticised or justified, since 'to talk of justifying them is to talk nonsense' (1940: 46). This is at once an historicist and a deflationary conception of metaphysics: talk of arbitrarily originating metaphysical schools is to be replaced with a piecemeal historical approach to 'constellations' of '*consupponible*' (1940: 66) presuppositions hanging together in any given historical period. Specifically, the metaphysician's job is ancillary (or at least supplemental) to that of the scientist, so 'what are misdescribed as A's "metaphysical doctrines" are nothing more than the results of A's attempt to discover what absolute presuppositions are made by scientists in his own time' (1940: 68). All metaphysicians do is to probe and uncover, after the fact, what the spontaneous philosophy of the scientists of their own age (or the past) is. Metaphysics is always an intellectual history of the present.

The unacknowledged reference, in the preceding paragraph, to Althusser's 1967 *Philosophy and the Spontaneous Philosophy of the Scientist* lecture (in Althusser 1990) is not casual. An analysis of Collingwood's thesis from a Bachelardian/Althusserian standpoint will allow me to focus on its most problematic shortcoming: the lack of a satisfactory account of how absolute presuppositions (or more correctly, constellations of them) are replaced with new ones. Collingwood was sufficiently historically nuanced (and enough of a Hegelian) to stress that '[w]here there is no strain there is no history' (1940: 75) but what he has to offer in the way of an explanation for the historical passage from one constellation of absolute presuppositions to another is hardly satisfying. Having proclaimed that a shift in absolute presuppositions is 'the most radical change a [human] can undergo, and it entails the abandonment of all [his or her] most firmly established habits and standards for thought and action', Collingwood offhandedly proceeds by rhetorically asking

> [w]hy, asks my friend, do such changes happen? Briefly, because the absolute presuppositions of any given society, at any given phase of its history, form a structure which is subject to 'strains'...of greater or less intensity, which are 'taken up'...in various ways, but never annihilated. If the strains are too great, the structure collapses and is replaced by another, which will be a modification of the old with the destructive strain removed; a modification not consciously devised but created by a process of unconscious thought. (1940: 48)

I am not sure how Collingwood's friend might be appeased by this allegorical explanation (an *obiter dictum* relegated to a footnote of his *Essay*). Of course, confronted with a more insistent questioning, Collingwood would simply retreat behind his *definition* of absolute presupposition as an unjustifiable background conviction and dismiss it as an invitation to pseudo-metaphysics. But this stance is, quite simply, untenable because it relinquishes the progress of history (and in fact the progress of science) to a randomly evolving flow propelled by a-rational (if not downright irrational) unconscious tensions. The metaphysician/historian would then be forced simply to accept the constellation of absolute presuppositions under inquiry as a brute fact, internally consistent (if subject to internal strains) but with little or no connection with what came before or will come after – incommensurable, to borrow the infamous Kuhnian term. Anyone interested in the development of science and its presuppositions (convinced that there is nothing like completely presupposition-less knowledge), and indeed sharing Collingwood's opinion that 'different sets of absolute presuppositions correspond not only with differences in the structure of what is generally called scientific thought but with differences in the entire fabric of civilisation' (1940: 72), cannot but find Collingwood's stance excessively unambitious and indicative of an historicist metaphysics flattened to the stale enumeration of one damn presupposition after another. In other words, whilst necessary, this non-committal kind of historicist descriptive metaphysics[25] can and should be accompanied by both a stronger interpretative key (looking for patterns of progress) and, as I will argue below, a more creative, revisionary metaphysics.

Bachelard's approach to the history of science can work as a corrective to Collingwood.[26] Like Collingwood, Bachelard adopts a psychoanalytic vocabulary to describe the development of the scientific mind and knowledge, stressing the necessity of employing a 'psychoanalysis of reason' (2002: 29) in order to unearth unconscious patterns of thought at work in the scientist's mind. But *unlike* Collingwood he does so by operating on this historical material in the attempt to trace a discontinuous process of rational development, without for that reason construing it as the linear, seamless development of absolute objectivity (the kind of postitivist-inspired 'deductive metaphysics' that Collingwood wanted to attack).[27] In other words, the description of conceptual presuppositions must be accompanied by their constant questioning precisely to the extent that they seem to display a self-evident nature, and thus might constitute epistemological *obstacles* to new scientific knowledge. As Bachelard writes: '[a]n epistemological obstacle will encrust any knowledge that is not questioned. Intellectual habits that were once useful and healthy can, in the long run, hamper research' (2002: 25). Bachelard describes the scientific spirit as becoming explicit precisely through science's endlessly self-critical stance,[28] where this series of strains and ruptures, or epistemological breaks, between commonsensically obvious (ideological, as Althusser would have it) assumptions and counter-intuitive scientific novelty, and even between successive scientific frameworks, is the veritable mark of modern thought.[29] Epistemological obstacles are, in Collingwoodian terms, the residue of past absolute presuppositions which need to be exposed and rectified in order for scientific thought to progress,[30] and the study of theory-formation needs to account for the way in which such presuppositions, *qua* misleading images and metaphors (mostly drawn from the field of common-sense experience), constitute stumbling blocks for the formulation of new scientific ideas. So Bachelard writes that '[a] science that accepts images is, more than any other, a victim of metaphors. Consequently, the scientific mind must never cease to fight against images, against analogies, and against metaphors' (2002: 47). Note, however, that this passage can be deceptive: Bachelard is not *rejecting* the use of metaphors, but is warning against a supine acceptance of them. Metaphors are not resting places for the scientific mind but way-stations to be continually reinvented by the creative scientist.[31]

The mark of the scientific intellect, for Bachelard, is the endless dialectical struggle against the laziness of thought[32] (and it is no surprise that leftist intellectuals from Althusser onwards have fully exploited the Marxist semantic resonance of these terms). Creativity, Bachelard points out with Nietzschean insight, finds fertile grounds in strife and unsettlement: to be devoted to a rationalist *engagement* means to acknowledge how

> reason loves to be in danger. If in a new experience one does not gamble with one's reason, that experience is not worth being tried. The risk of reason, after all, must be total. That is its specific character.

> It is all or nothing... Indeed, what would I do with yet another experience that confirms what I already know and, by consequence, what I already am? Every real discovery determines a new method and dismisses a prior method. In other words, in the realm of thought, imprudence is a method... Those long-time gathered, patiently collated and avidly preserved knowledges are suspect. They bear the pernicious sign of prudence, of conformism, of constancy, of slowness. (1972: 11; my translation)[33]

The moral of this Collingwoodian–Bachelardian excursus is thus that (a realist) metaphysics can and should be at the same time historical and creative, epistemologically attentive to the ways in which present and past science tacitly assumes conceptual frameworks but daring to put new concepts to the test, in the attempt to make sense of science's deliveries – a creativity and a *conatus* of continuous self-critical innovation which should be part of the naturalist's creed.[34] It is possible, *pace* Collingwood, to *give* (and be *asked for*) reasons for the evolution and change in meta-scientific assumptions (hence offering a *non-dogmatic justification* for them) and it is therefore possible to evaluate their historical succession along a rational-progressive axis: as Tiles observes vis-à-vis Bachelard, 'this scientifically internal history of science is *necessarily evaluative*' (Tiles 1984: 13; emphasis addded).[35] Our metaphysics then, to quote A.W. Moore's eloquent description, 'consists in the protection, nurturing, adaptation, rejection, or replacement of some of our most general concepts, systems of classification, ways of thinking, and the like' (2012: 583).

3.5 The realist's task

Regarding naturalism my conclusion is that, first, naturalism is a philosophical or meta-scientific thesis (not a scientific one) and represents a more or less unitary set of presuppositions regarding reality, a set unified by an ontological commitment to immanence (there is nothing outside the causal order of the universe,[36] and even the unknown or unknowable can be known to be part of it). Second, what metaphysical speculation can do is neither overthrow naturalism nor correct the natural sciences but instead question the ways in which naturalist commitments are cashed out: naturalism, as I construe it, makes no pronouncement regarding the internal organisation of reality, *except* that any organisation it *might* have is a fully contingent one. All that it pronounces is 'this is all there is' leaving open the ontological task open of figuring out what 'there is' *is*, and the epistemological one of how we can get to know it, in both cases assigning primary but not incorrigible epistemic authority to the natural sciences, from physics to the neurosciences.[37] It seems to me, then, that several ontological options are open to the naturalist, and it is the task of philosophy both to scan for possible sources of theoretical bias in scientific ideas *and* create the concepts necessary in order to talk about alternative routes beyond these obstacles. The continuous feedback loop is then between a science (fundamental physics) that forces metaphysics to revise its conceptual toolbox[38] and a metaphysics that helps to make sense of science's discoveries by inputting new concepts and offering revised meta-scientific frameworks capable of re-orienting methodologies and correcting assumptions.[39] Daniel Dennett has famously argued for the indispensability of philosophy, observing that

> [s]cientists sometimes deceive themselves into thinking that philosophical ideas are only, at best, decorations or parasitic commentaries on the hard, objective triumphs of science, and that they themselves are immune to the confusions that philosophers devote their lives to dissolving. But there is no such thing as philosophy-free science; there is only science whose philosophical baggage is taken on board without examination. (1996: 121)[40]

This, of course, does not mean that the philosopher is supposed to overlook every stage of the scientist's work, from theory-formation to experimental set-ups. In fact, this kind of philosophical work can only be applied on the larger, most *general* scale: proposing different, large-scale ways of conceptually framing the motley array of scientific theories and results.[41] As Michael Friedman has argued, defending his neo-Kantian approach to historically relativised *a priori* principles, philosophy's 'peculiar role is precisely to articulate and stimulate new possibilities, at the meta-scientific level' since '[w]e never know in advance what new paradigms (and philosophical meta-paradigms) might be needed at a given moment of revolutionary science' (2001: 24).

The most delicate (since most general) set of assumptions or meta-paradigms the realist philosopher of science should look out for (or at least the most important ones vis-à-vis the anti-theological project of this book) are those related to the *intelligibility* of the *ordered* universe, those principles offering transcendental support for the desire to place singular facts on a large-scale tapestry of meaningful connections, to see reality given to us with an intrinsic semantics. These are not now-superseded early modern assumptions, perhaps interesting for the historian of ideas but irrelevant to the contemporary philosopher of science. As recently as the 1920s A.N. Whitehead clearly spells them out when arguing that 'there can be no living science unless there is a widespread instinctive conviction in the existence of an *Order of Things*, and, in particular, of an *Order of Nature*', proceeding then to describe this conviction as 'a faith which is impervious to the demand for a consistent rationality' (1948: 4) and 'an un-conscious derivative from medieval theology' (1948: 14). Stephen Toulmin repeatedly stressed the importance of recognising this particular assumption, arguing that

> in studying the development of scientific ideas, we must always look out for the ideals and paradigms [humans] rely on to make Nature intelligible. *Science progresses, not by recognizing the truth of new observations alone, but by making sense of them.* To this task of interpretation we bring principles of regularity, conceptions of natural order, paradigms, ideals, or what-you-will: intellectual patterns which define the range of things we can accept (in Copernicus' phrase) as 'sufficiently absolute and pleasing to the mind'. An explanation, to be acceptable, must demonstrate that the happenings under investigation are special cases or complex combinations of our fundamental intelligible types. (1961: 81; emphasis added)

It is indubitable that human beings cannot do without explanatory frameworks, conceptual structures and (regulative) ideals employed to organise experience, co-ordinate epistemic efforts, and make sense of the (increasingly vast) output of data originating from scientific experimentation. Science would not progress (or be of much interest) if limited to a mere collection of disjointed brute facts. It is also the case that there is nothing like a pure, theory-free perception or a theoretical construction unalloyed by background assumptions, even in the seemingly straightforward case of perception of physical objects, as there is no extra-conceptual access to reality. But this shouldn't cast us into sceptical despair since, as Sellars put it, '[o]ur classifications of physical objects can become more complex and sophisticated [and] [o]ld pigeon holes can be subdivided, change their shape, and even disappear' (1976: 318).[42] It is thus the realist's responsibility to ensure that the historical change of such conceptual structures is conducive to their *improvement*, that is, allows an increasingly detailed, more accurate grasp of the reality they are applied to (starting by replacing the concepts of commonsense experience with those delivered by science). As Lakatos put it '[s]cience teaches us not to respect any given conceptual-linguistic framework lest it should turn into a conceptual prison' (1976: 93, n. 1). In other words, while we just cannot do without them, there are ways to adjudicate between competing presuppositions[43] and the necessity of *making sense* of the universe (to satisfy our theoretical curiosity and our practical necessities) should not be confused with the comfortable but intellectually lazy habit of *decoding sense* from its senseless, normatively inert, causal structure.

Perhaps a distinction would be appropriate. Conceptual influences on our inquiry into reality can be classified in two broad groups. First, there is the network of *concepts* and inference rules, developed in the span of our lifetime or selected and inherited from millions of years of reproductive-success-driven evolution, which influence and constrain our perception: these are akin to the Kantian categories and permit the successful navigation in our world of everyday experience. These concepts are at once the hardest to disengage from in our common experience and the easiest to bracket or modify during methodologically regulated scientific inquiry: consider the case of perceptual illusions which persist even when we possess a rational explanation for the illusion itself (the Ames room is a good example among many). The other kinds of influence, which we might call *principles* (closer to the Collingwoodian absolute presuppositions, or even Baconian idols), are meta-experiential concepts which systematise the *totality* of our experiences, and as such are often very hard even to *identify*. The discovery and examination of the former ones are (for the most part) the occupation of the epistemologist, the philosopher of perception or, nowadays and increasingly so in the future, of the neuroscientist and cognitive psychologist. The uncovering of the latter ones is the concern of the intellectual historian and the Collingwoodian-inspired metaphysician. While humans, as cognitively finite beings, will never be able to disengage fully from many perception-structuring concepts (which need not be an obstacle for the realist: once again here a Sellarsian argument can be marshalled in favour of the regulative ideal of a desirable fusion of the images wherein 'the methodologically important dualism of observation and theoretical frameworks would be transcended, and the world of theory and the world of observation would be one' (1974: 453)), it is my argument that we need to be relentlessly self-critical regarding the provenance, articulation and legitimacy of those *principles* (such as the principle of sufficient reason, the belief in the simplicity, nomological uniformity and intrinsic organisation of the universe, and so on) that *can* be revised and corrected, being primarily a product of our (relatively) recent intellectual history.

If scientific realism, as I argued above, is a thesis about the universe which centrally entails a rejection of anthropocentric forms of thought and an understanding of reality as impervious to (common-sensical, manifest image-based) human cognitive efforts, then one of the primary concerns of the scientific realist, aided in her task by the historian's *cognitive-historical* analysis,[44] must be that of purifying her conceptual framework (and any set of super-empirical criteria guiding the process of scientific theory-choice)[45] of the deeply ingrained and unconscious assumptions of the intelligibility, purposiveness and rational order of the universe – yet without forfeiting all epistemic ambitions and retreating into a fideism of wonder.

Above all, the target of the realist should be whatever is left over of that powerful network of beliefs running throughout Western philosophy which was memorably examined by Arthur Lovejoy in his *The Great Chain of Being*. This was an idea to be epistemologically cashed out in 'a faith, implicit or explicit, that the universe is a rational order, in the sense that there is nothing arbitrary, fortuitous, haphazard in its constitution' (2001 [1936]: 327). According to Lovejoy

> [i]n so far as the world was conceived in this fashion, it seemed a coherent, luminous, intellectually secure and dependable world, in which the mind of [humans] could go about its business of seeking an understanding of things in full confidence; and empirical science, since it was acquainted in advance with the fundamental principles with which the facts must, in the end, accord, and was provided with a sort of diagram of the general pattern of the universe, could know in outline what to expect and even anticipate particular disclosures of actual observation. (2001 [1936]: 328)

Against this failed dream, what a fully naturalised realism must aim at is the jettisoning of the nexus of intelligibility, reason and meaning[46] operating at either the metaphysical or phenomenological level, while being devoted to offering an account of the relation between reason and reality. A jettisoning, then, that need not amount to a principled *reversal*, which would risk leading straight to a reactive mystical quietism,[47] a forfeiting of reason's ambitions. With his characteristic penchant for poignant and compressed rhetorical reversals, Brassier memorably encapsulates the realist creed in his lapidary claim that 'nature is not reasonable, and reason is not natural, yet nature's unreasonableness is not unintelligible, just as reason's unnaturalness is not supernatural' (2012).

The scientific realist project, then, should move between the Scylla of unquestioned belief in the power of an idealised and cognitively omnipotent rational scientific inquiry and the Charybdis of a defeatist sense of epistemological entrapment within a hostile and mysterious universe – between, in other terms, the (rationalist) infinity of reason and the (naturalised) finitude of humanity.

In the next chapter I will introduce and explore what I take to be the most promising metaphysical picture to defend in accordance with these realist commitments, one that will allow me to further pursue the naturalisation of Badiou's mathematical ontology.

4
Structural Realisms

In the previous chapter, I spelt out what I believe the commitments of any realist metaphysics should be, articulated through the two directives of historicism and naturalism. It is time now to examine in detail the specific form of scientific realism that can move towards a naturalist revision of Badiou's project, that is structural realism. In order to do that, a modicum of historical background is in order.

4.1 Scientific realism, its abductive defence…and its discontents

Scientific realism reacquired philosophical currency after the decline of the logical positivist project and its attempted overthrow (*Ueberwindung*), through logical analysis, of metaphysics in general and of non-empirically verifiable theoretical postulates in particular.[1] The latter, the Vienna Circle philosophers argued, were to be explained away. Scientific theories, in their so-called 'received view', were supposed to be modelled on formal systems of uninterpreted predicates and inference rules capable – through the application of adequate correspondence rules (or 'bridge laws') allowing for a specific interpretation – of deriving consequences regarding the behaviour of (and making inductive generalisations about) objects in the observational framework, the sphere of the empirically given.[2] Privileging syntax over semantics, the basic units of theories were taken to be strictly empirically verifiable 'protocol statements' or 'observation reports', while theoretical terms naming unobservable objects were dismissed as just words without a referent.[3] As stated in the 1929 Vienna Circle

Wissenschaftlichen Weltauffassung manifesto: 'in science there are no "depths"; there is surface everywhere' (Hahn et al. in Neurath and Cohen 1973: 306).[4]

The logical positivist project of rational reconstruction of scientific theories, hindered by an overtly rigid verificationist criterion of meaning according to which 'the meaning of a proposition can be given only by giving the rules of its verification in experience' (Schlick 1936: 342), failed to pay the expected philosophical dividends. Reacting against what came to be perceived as a stagnant project, around the 1960s a number of philosophers started questioning the ability of any such ontologically non-committal position to offer an adequate explanation of scientific phenomena, particularly targeting the alleged failure of theoretical terms to pick out real features of reality. Perhaps ontological questions had been too hastily dismissed as 'pseudo-questions' after all.[5] Realists reconsidered causal models of explanation,[6] as replacing deductive-nomological ones: the explanandum is to be placed in a causal network, and causal responsibility must be assigned to theoretical entities. An *etiological explanation* – linking causes to effects – requires an ontological commitment to unobservable entities: as Sellars put it, 'to have good reason for holding a theory is *ipso facto* to have good reason for holding that the entities postulated by the theory exist' (1963: 91).[7]

The realists argue that anti-realist (phenomenalist or instrumentalist) interpretations of scientific terms as mere placeholders, allowing the scientist to achieve her predictive goals but indexing no deeper reality beyond their use in the theoretical vocabulary, cannot but lead to the belief in a sort of '*cosmic coincidence*' to use J.J. Smart's (1963: 39) well-known phrase. No explanation could be offered for the constant surprise of observed phenomena regularly succeeding each other. 'On the other hand', Smart observed, 'if we interpret a theory in a realist way, then we have no need for such a cosmic coincidence: it is not surprising that galvanometers and cloud chambers behave in the sort of way they do for if there really are electrons etc., this is just what we should expect' (1963: 39).[8] Roy Bhaskar similarly championed a transcendental realism against 'actualism' (the empiricist view which recognises as real only nomologically related events and states of affairs object of actual or possible experience), arguing that 'the ultimate objects of scientific understanding are neither patterns or events nor models but the things that produce and the mechanisms that generate the flux of phenomena of the world' (2008: 66).

The form of inferential reasoning buttressing these arguments goes by the various names of abduction, ampliative inference, retroduction or inference to the best explanation (IBE).[9] Broken down to steps, the inferential pattern has the following structure:

1. enumeration of relevant observed data;
2. postulation of a number of possible explanations which would be able to explain them;
3. choice of the best (according to the theoretical virtues of simplicity, explanatory power, coherence with known facts, fruitfulness, general consilience with the rest of the theoretical edifice…) of these explanations;
4. conclusion that the previously unknown phenomena and entities invoked as the best explanation of the observed facts are in fact real.

The peculiarity of abductive reasoning, then, is the stress put on explanatory considerations: abduction is an *ampliative* inferential process to the extent that at its conclusion the reasoner possesses *more* information than she did at the beginning:[10] '[i]n ampliative inference… a risk is taken, a venture is made beyond the strict limits of the original evidence' (McMullin in Hilgevoord 1994: 81).[11] The additional information concerns entities and processes causally responsible for the *explanandum*. Many, if not most, of our inferential practices are guided by explanatory considerations and follow this pattern: from our everyday stopping to think where we can possibly have left our mobile phone to scientists attempting to find an explanation for the unusual readings on their laboratory apparatus. Scientific realists support IBE for its specific role in scientific inquiry:[12] IBE is, as per McMullin's 1992 title, *The Inference that Makes Science*.[13]

The theorisation of IBE has a long history, often traced back to Charles Sanders Peirce's reflections scientific reasoning in the late nineteenth/early twentieth century and in particular to his coinage of the term 'abduction'.[14] Peirce distinguished three elementary kinds of reasoning – deduction, induction and abduction – but observed that

> [i]t [induction] never can originate any idea whatever. No more can deduction. *All the ideas of science come to it by the way of Abduction.* Abduction consists in studying facts and devising a theory to explain them. Its only justification is that if we are ever to understand things at all, it must be in that way. (*CP* 5.145; emphasis added)

And, even more explicitly, he suggests that 'every single item of scientific theory which stands established today has been due to Abduction' (*CP* 5.172). From Peirce onwards, the central epistemic trait of abduction, then, is the *creation* and the 'probational adoption of' explanatory hypotheses (*CP* 2.96). Unlike deduction and induction (but also unlike *prediction*)[15] it is an inferential avenue which puts a premium on the conceptual inventiveness of the active mind of the inquirer: faced with a surprising state of affairs, the rational agent seeking understanding via explanation attempts to revise or update her conceptual schemes via the postulation of new entities or phenomena capable of accounting for it. Abduction, however, 'merely suggests that something *may be*' (*CP* 5.171), and the process of selection and confirmation of the best explanatory hypothesis requires both empirical testing and (when this proves impossible) evaluation according to a range of theoretical virtues. That is to say, in Bachelardian spirit, that abduction is no context-independent rationalist inferential method, but needs to be developed in a constant dialectic *with* experimental practice.

In the following decades, scientific realism gained progressively more traction in the philosophical community, and was defended through additional arguments. Broadly speaking, the specific ampliative/abductive argument for the indispensability of theoretical entities to flesh out causal stories was supplemented by a meta-scientific argument, focusing not only on the correct ontological commitments of any given theory, but on the success of the scientific enterprise as a whole. This success (both in terms of theoretical predictions and experimental manipulation), the realists argued, can only be fully explained by the meta-induction that the natural sciences, in their most mature forms, manage to grasp – to at least a high degree of approximation – the truth about (different facets of) the physical world. This became known as the 'no miracles argument', after Putnam's remark that '[t]he positive argument for realism is that it is the only philosophy that doesn't make the success of science a miracle' (1975: 73). The large-scale predictive success of the sciences, the argument goes, gives us reason to believe that the postulational strategy picks up real entities.[16]

These IBE-based lines of argument, however, failed to convince those who opposed to them both a specific and a meta-scientific counter-argument.[17] The former, generally traced back to Pierre Duhem[18] and known as the Quine–Duhem thesis (after the former's reactivation of the latter's intuition) argues that no single theory can be crowned as a uniquely valid explanation of a given phenomenon. When reference is made to unobservable entities and causal processes, any phenomenally manifest event can be explained by an endless list of theories referring to the unobservable micro-level – theories which, crucially, are all empirically equivalent. This results in the *underdetermination of theory by data*: no amount of experimentally gathered data should convince the scientist, so the argument goes, of the correctness of the one theory used to operate an explanatory synthesis on them. For any theory T it will always be possible to devise a theory T′ such that the empirical consequences of the two will be phenomenally indistinguishable.

The meta-scientific argument, popularised above all by Larry Laudan in a renowned paper (in Curd and Cover 1998) and commonly known as the *pessimistic meta-induction* (PMI), states that the historically minded philosopher cannot unquestionably accept realism once due consideration is given to all those past theories which, (in today's light and from the realist's own point of view), were failures due to their reference to actually non-existing entities or properties (the usual suspects here being crystalline spheres, caloric, phlogiston or the ether). In certain cases, even where syntactic continuity occurred, the semantic content of the theory's terms would change beyond recognition (think of the different meaning of the term 'electron' in classical and quantum physics). Laudan's most far-reaching conclusion, meant as a constructive criticism of an overtly optimistic realism, was that 'realism *cannot,* even by its own lights, explain the success of [these] many theories' and that

> [t]he inescapable conclusion is that insofar as many realists are concerned with explaining how science works and with assessing the adequacy of their epistemology by that standard, they have thus far failed to explain very much. (In Curd and Cover 1998: 1131–2)

This line of argument can be generalised even further and, giving it a Kuhnian slant, defined as the 'argument from scientific revolutions': how can the realist support the thesis of a rational progress of science when history shows that mutually incompatible (or even logically inconsistent) scientific theories have succeeded each other in haphazard, unregulated fashion?[19]

4.2 Epistemic structural realism

Striving to reconcile the most convincing elements in the stances of both realists (the no-miracles argument) and anti-realists (the PMI and historical acknowledgement of theory-change),[20] John Worrall[21] defended the approximate truth of mature scientific theories by reviving a long-standing tradition of *structuralism* in the philosophy of science.[22] Retrospectively, it can be said that Worrall's paper laid the grounds for the contemporary research programme in structural realism, increasingly prominent in the last 15 years and today *the* most hotly debated form of scientific realism in the literature.[23] However, the structuralist approach – minimally interpreted as a methodological prioritisation of structural features of a domain of inquiry[24] – need not entail a metaphysical stance: while my focus will be on structural *realism*, empiricist or transcendental idealist forms of structuralism can be found throughout its historical evolution (early structuralist stances can be traced back at least to the 1910s) all the way to the present.[25]

Worrall, like Laudan, believed that realist accounts of science have thus far failed to offer proof for scientific theories' ability to offer an increasingly accurate approximation of reality and that indeed inescapable historical considerations pose the issue of semantic discontinuity as a threat that standard scientific realism cannot meet (a student under Lakatos, Worrall inherited his historicist scruples). Worrall thus recuperates a position most explicitly defended by Henri Poincaré in the early twentieth century, according to which the evolution of scientific theories is best understood as manifesting historical continuity not at the level of their referents (the unobservable entities referred to by the theoretical terms) but at that of the *structural relations* obtaining between unobservables. This position allows Worrall to fend off Laudan's (or Kuhn's) objections by preserving a robust sense of progressive *cognitive continuity* in science, while discarding the problematic modification of ontological commitments from theory to newer theory. Entities, in Kantian fashion, are considered by Poincaré and Worrall alike to be inaccessible to us in their true nature: what we can (and should) be realists about are the knowable relations between these not-fully-knowable (at least not in their non-relational properties) entities. And, crucially, the structure constituted by these relations emerges as intelligible to us within the mathematical formalisms (in Worrall's and Poincaré's examples, mostly *equations*) that we employ in successful theories.[26] So Poincaré argued that to vindicate scientific progress in the face of the apparent 'bankruptcy of science' (roughly, his term for the conclusion of the PMI), we should observe that

> equations express relations, and if the equations remain true, it is because the relations preserve their reality. They teach us now, as they did then, that there is such and such a relation between this thing and that; only, the something which we then called motion, we now

call electric current. But these are merely names of the images we substituted for the real objects which Nature will hide for ever from our eyes. The true relations between these real objects are the only reality we can attain, and the sole condition is that the same relations shall exist between these objects as between the images we are forced to put in their place. If the relations are known to us, what does it matter if we think it convenient to replace one image by another? (1905: 161)[27]

To construe science as aiming at the true nature of things is, Poincaré argues, a profound misapprehension of the very scope of knowledge:

> not only science can not teach us the nature of things; but nothing is capable of teaching it to us and if any god knew it, he could not find words to express it. Not only can we not divine the response, but if it were given to us, we could understand nothing of it; I ask myself even whether we really understand the question. (1958: 138)[28]

Whilst not explicitly guided by such momentous epistemological convictions, Worrall follows Poincaré's lead and takes as his main case study the continuity in mathematical structure detectable in the equations employed first by Augustin-Jean Fresnel first and later by James Clerk Maxwell in the attempt to offer an explanation for the phenomenon of propagation of light (similar sets of equations referring to the behaviour of radically different postulated entities: an elastic solid aether for Fresnel, an electromagnetic field for Maxwell). Worrall echoes Poincaré's remarks by arguing that

> [t]he rule in the history of physics seems to be that, whenever a theory replaces a predecessor, which has however itself enjoyed genuine predictive success, the 'correspondence principle' applies. This requires the mathematical equations of the old theory to reemerge as limiting cases of the mathematical equations of the new. (1989: 120)[29]

Thus the structuralist 'insists that it is a mistake to think that we can ever "understand" the nature of the basic furniture of the universe' (1989: 122). Note, then, that this kind of structural realism amounts to a less sanguine form of realism, a *restricted* realism professing a history-induced Kantian humility[30] allowing for, at best, *indirect* knowledge of unobservable entities/things in themselves[31] (while being as committed as the standard realist to our knowledge of both structural and non-structural features of observables). Worrall deflates the question of *conceptual change* (the different interpretations historically given to theoretical terms) by restricting the realist's focus on the *preservation of structure*. Unlike Laudan's own interpretation of the Fresnel/Maxwell dialectics, Worrall's purports to put forward an '*optimistic* induction ... concerning the discovery of *mathematical structure* rather than individual ontology' (1994: 336).

4.3 Ontic structural realism

With what could be compared to a post-Kantian move – turning an epistemic constraint into an ontological insight[32] – a new wave of structural realists (initially spearheaded by James Ladyman and Steven French)[33] argued, in the wake of Worrall's rehabilitation of structuralism, for a different interpretation of the manifest continuity of mathematical structures across theories, defending an *ontic* form of structural realism (OSR) against (what became known as) Worrall's *epistemic* structural realism (ESR). ESR, a half-way compromise between scientific realism and anti-realist instrumentalism restricting our epistemic ambitions to the knowledge of structures and curtailing that of individuals, offers too little ontological clout for the scientific realist.[34] The insight concerning preservation of structure can be retained, while unknowable objects can be dispensed with. OSR, then, ontologises Worrall's structuralism, arguing that from the fact that structural explanations are enough for the descriptive and predictive success of science it follows that there is *nothing to know but structures*.[35] This amounts to nothing less than a 'shift to a different ontological basis altogether' (Ladyman 1998: 420), the conceptual and ontological demotion of objects and monadic relations (properties) to derived existence, and the preservation of relations as ontologically primary components of reality.

Given that the metaphysical interpretation of elementary particles described by our best microphysical theories is underdetermined by the formalism of the theory itself – as in the case of quantum mechanics, where 'particles' seem to violate the Leibnizian Principle of the Identity of Indiscernibles (PII)[36] and can be ontologically interpreted as either individuals or as non-individuals[37] – OSR argues that what we can and do know about are the structure(s) of reality. That is because, the argument goes on, no inaccessible domain of entities/individuals/intrinsic properties lies beyond structural descriptions so that different interpretations of these 'individuals' would be nothing but different (and equally misguided) representations of the same underlying structure.

Interpretative issues arising from metaphysical underdetermination are moot: there are *neither* individuals *nor* non-individuals, only structures.[38] In terms of basic ontological commitments, then, ESR preserves intrinsically existing entities considering them indispensable for the instantiation of structures, while OSR takes structures as free-standing realities. Not only, as Worrall's ESR holds, does there obtain a one-to-one, structure-preserving mapping or isomorphism between physical reality and its mathematical model, but this isomorphism completely exhausts the semantic problem of how to relate the mathematical and the physical, since there is no more to the physical (no hidden, 'natures' or non-relational properties) than appears in the mathematical. So while in Worrall's approach the threat of underdetermination of theory by data is only side-stepped (no direct reference is made to unobservable entities), OSR aims at completely defusing it since all there is to know about unobservables is presented by the mathematically structured theory. Against PMI arguments, on the other hand, OSR claims (to retrieve Worrall's example) that it is not the case that the electromagnetic field exists (for current best physics) while the aether does not, but simply that *neither* exists over and above (or below) the equations describing their properties and behaviour.

According to OSRists, ESR fails to contribute enough to scientific realism, and a metaphysical interpretation of structuralism is necessary: our best present physics (particularly quantum-field theory [QFT])[39] strongly undermines any conception of individual 'entities', assigning them, at best, a heuristic, instrumental or contextual coherence. As Ladyman puts it, 'questions of individuality simply do not arise' (1998: 420), and if they do it is already by making the wrong ontological assumptions, since 'simply abandoning individuality won't necessarily yield a non-object oriented ontology' (French 2010: 94).[40] Current forms of OSR, taking structures 'to be primitive and ontologically subsistent' (Ladyman 1998: 420), achieve unprecedented consilience by being the only position at once supported by current best science and capable of defusing both the anti-realist challenges of underdetermination and theory-change.[41]

The elementary particles described by physics, then, are nothing but the ripples/nodes in an underlying field/structure, or, in group-theoretical terms, they are 'just sets of quantities that are invariant under the symmetry groups of particle physics' (1998: 421). Realist metaphysicians who cling to an individual-based ontology 'haven't learnt the lessons of quantum physics' (French and Ladyman 2003a: 38), display their 'inability to escape the manifest image' (Ladyman 2009: 140) and ultimately are 'people who know just a bit of superficial science [and] are comfortable with thinking about a world made out of ultimate little things and collisions amongst them' (Ladyman and Ross 2007: 23).[42] As Jonathan Schaffer (not himself an OSRist, but similarly opposed to the notion of a 'fundamental level') sharply puts it, '[t]he claim that a complete microphysics must be a theory of particles, much less of discrete entities at all, is just an article of faith' (2003: 504).[43]

The version of OSR I am delineating here (as developed by Ladyman, French and Ross) should be further qualified by contrasting it with other variants available to the metaphysical structuralist. Frigg and Votsis (2011), distinguish between a radical ontic structural realism (ROSR) and an eliminative ontic structural realism (EOSR):[44] the former plainly claims that *all that there is* is (abstract) structure while the latter argues that only structure is *ontologically fundamental*. While both positions hold that structure is just relations without relata, the main difference lies in the different understanding of 'relations'. So

> [w]hile ROSR insists on an extensional treatment of relations that is standard in mathematical logic, EOSR allows for relations that have intensions; that is, fundamental relations can be relations like 'being larger than' rather than only 'standing in a transitive relation'. In other words, relations are interpreted. (2011: 262)

Ladyman and French, in this classification, are defending EOSR, ROSR being considered untenable by reason of its inability to account for physical instantiations of structure. However, even the eliminative stance is often considered more revisionary of classical metaphysics than required to account for scientific results, or unfit to offer a coherent scientific realism.[45] The main objections raised against OSR stem from a general scepticism regarding the very intelligibility of an object-less metaphysics and (consequently) its adequacy as an ontological grounding for the observable physical world. In particular, critics have tended to single out the metaphysical impossibility that relations should subsist prior to relata and the lack of causal efficacy of relata-(object-)less structures. Stathis Psillos denounces OSR as holding the

> wrong ontological thesis that structures require no individuals in order to exist and the wrong epistemic thesis that they can be known independently of (some, but not any in particular, set of) individuals which instantiate them... Unless we buy into some problematic metaphysical thesis which somehow 'constructs' the individuals out

> of relations, the world we live in (and science cares about) is made of individuals, properties and their relations. (2009: 135)

He further further insists that

> [p]laces in structures and formal relations do not cause anything at all. It's the 'fillers' of the places and concrete ... relations that do ... The phenomena are able to give 'content' to a structure precisely because they are *not* themselves structure. (2009:141)

Busch agrees since 'the very idea of structure presupposes some elements that go together to make up that structure. A relation might take *anything* as its relata, but it always takes *something*' (2003: 213). Anjan Chakravartty argues that while relations can be said to hold a *conceptual* dependence to relata ('it is part of the very concept of a concrete relation that it relate *something*' (2007:77)) it is the stronger notion of *causal* dependence that creates a more significant obstacle for OSR, since it is only thanks to properties-bearing objects that any form of change can be accounted for: without the ontological 'oomph' of objects, there can be no explanation for the transformation of a given structure into another. Hence his own compromise position, labelled *semi-realism*, is derived from the consideration that OSR must covertly make reference to traditional entity realism to make sense of scientific phenomena.[46] In the context of a debate with Ladyman and French, Cao[47] objects to their dissolution of entities in mathematical structures, arguing, first, that physical structures involve qualitative elements that cannot be captured by a mathematical structure; second, that, unlike a mathematical structure, the meaning of a physical structure needs to be drawn from relata which ontologically precede relations; and, third and most importantly, that 'the crucial criterion for judging ontological priority is causal efficacy, which provides a basis for explanation and prediction' (2003b: 67), a causal efficacy that cannot be satisfactorily predicated of mere mathematical structures. Finally, Esfeld and Lam put the case well when they assert that

> [t]he main objection against the ontic structural realism of French and Ladyman ... is that it is too parsimonious: relations presuppose relata, that is, objects between which the relations obtain and of which they are predicated. More precisely, a Platonist may maintain that relations as such exist as abstract structures, that is, abstract entities that are universals. However, when it comes to the physical world, the

> point at issue are [sic] concrete relations that are instantiated in the physical world and that hence are particulars in contrast to universals. For the relations to be instantiated, there has to be something that instantiates them, that is, something that stands in the relations. In brief, the objection is that in eliminating objects [OSR] is not intelligible as a theory of the physical world: it is not able to distinguish between structure as an abstract entity and the structure of the physical world. (2008: 31)[48]

What response can OSR offer to these arguments without reintroducing some conception of object in its ontology?[49] French and Ladyman believe that their position need not be threatened by the problems of causation and of relata-less relations, and reprimand their critics' inability to make a radical enough conceptual leap into a fully de-objectified metaphysics and for offering question-begging challenges, grounded in traditional, non-naturalised metaphysical discussions. Their arguments have a strongly naturalist and reductionist flavour: macroscopic causal processes displaying non structure-preserving change are not a problem for the OSRists to the extent that the latter merely 'piggybacks on the physicalist's reduction of such processes in terms of ultimately quantum processes and then insists that those have to be understood in structuralist terms' (in Bokulich and Bokulich 2011: 36). Asked about some account of causation presupposing some*thing* undergoing change, some substance or object, the OSRist responds that

> what we have is the world structure with a particular configuration, if you like, or family of relations at one time and a different configuration, or a different arrangement or family of relations at either earlier or later times. That is all that is needed... [W]e are not convinced that the requirement that causality have some 'active' component requires a metaphysics of objects and properties. (In Bokulich and Bokulich 2011: 36, 38)[50]

As for the anxiety about relations without relata, the OSRists' defence pivots upon their naturalism and points out their critics' reliance on obsolescent metaphysical tenets and conceptual schemes, betraying an unwillingness to take on board the lessons of contemporary physics. Ladyman and Ross assert that

> [b]oth QM and relativity theory teach us that the nature of space, time, and matter raises profound challenges for a metaphysics that

> describes the world as composed of self-subsistent individuals. In so far as quantum particles and spacetime points are individuals, facts about their identity and diversity are not intrinsic to them but rather are determined by the relational structures into which they enter. (2007: 151)

They also trenchantly surmise that

> [g]iven that there is no a priori way of demonstrating that the world must be composed of individuals with intrinsic natures, and given that our best physics puts severe pressure on such a view, the PNC[51] dictates that we reject the idea altogether. (2007: 154)

QED. Science first, traditional metaphysical quandaries and commonsense categories second. Besides, in most cases (that is, in all those contexts which are not that of fundamental particle physics), relata *can* be heuristically identified, since the OSRists are 'not denying that we can continue to refer to such objects as a *façon de parler*, only that these must themselves be regarded as decomposable into or reconceptualised as nodes in a structure or whatever' (French and Ladyman in Bokulich and Bokulich 2011: 38). This position, importantly, commits the realist to the rejection of the notion of a *fundamental* ontological level – encapsulated in the often used slogan 'it's relations all the way down' – while allowing for meaningful talk of 'objects' with 'intrinsic' properties[52] at different scales. The core ontological criteria of structuralism are precisely *anti-foundationalism* and *mutual individuation*: the 'fundamental' elements of reality do not enjoy 'fundamental' self-existence, and they can only be assigned a conventional interdependent existence, fully dissolved in their relations.

An interesting defence of the logical priority of relations and the dispensability of relata is offered by Floridi. Acknowledging the difficulty of demonstrating how essential properties of relata might depend on some internal relation, Floridi argues that 'there is a significant exception, a case that is ontologically more fundamental than the case in which the essence of the *relata* is in question. This is the (internal) relation of *difference*, which constitutes its *relata*' (2008: 234). He proceeds by explaining that

> [t]he relation of difference seems a precondition for any other relation and hence for any process of knowledge. *Relata* as merely *differentiated* entities and nothing else (at least not yet) are possible only

> because there is a relation of initial, ontological difference in the first place. (2008: 234)

He thus proposes a basic ontological principle of *difference* as making intelligible the notion that relata are logically preceded by a relation, so that '[d]ifference is our *Ur*-relation' (2008: 235). Even though Floridi ends up defending a more moderate 'package hypothesis', where relations and relata are equiprimordial ontological categories, this strategy for defending OSR's 'full plausibility' (2008: 236) is mostly noteworthy, in the context of my 'cross-traditional' exposition, for its conspicuous resonances with the general strategy of French (Saussurrean) structuralism and its Derridean deconstruction,[53]

The adoption of a structuralist metaphysics, no matter how eliminativist, does not force, the OSRist insists, the absurd relinquishing of our reliance on manifest, observable objects, nor indeed need it deter scientists from their search for fundamental, individual particles. To do so would mean pushing the selective scepticism which guided Worrall's initial move to an inane rejection of any phenomenologically substantial entity, and turn the structuralist into a modern-day Pyrrho, having to be saved by his friends from walking in front of an oncoming, and all-too-substantial, vehicle. The structuralist wants to *reconceptualise* entities in structural terms, not to stop believing in them: OSR is a *realism* precisely to the extent that it describes our cognitive effort as pursuing (through scientific means) a mind-independent reality which enjoins a shift in our conceptual schemes and embrace a structuralist framework.[54] Our semantics should also be reconceived structurally, mindful of the referential distortion that inevitably occurs whenever formal (mathematical) structures are translated into natural languages. As Ladyman and Ross clearly explain,

> [t]o the question 'Are there really protons?' the answer is 'Yes', because theories in which protons are elements characterize real structure; but it does not follow from this that the world is partly composed out of individual pieces that intrinsically bear properties corresponding to predicates of the word 'proton' as it occurs in natural language paraphrases of theories in mathematical physics. (In Ladyman et al. 2013: 127)

The adoption of the Sellarsian trope of a stereoscopic integration of manifest and scientific images is once again expedient: the new structuralist (scientific) conceptual framework describes the actual ontological make-up of the manifest image but cannot, in one clean sweep, replace its methodological priority (cf. Sellars 1963: 20). Indeed, we should not shy away from the naturalist imperative of *updating* the Sellarsian terminology, expanding his reference to the 'swirl of physical particles, forces, and fields' (1963: 20) populating the scientific image of the 1960s with an image of object-less structures suggested by our current physics. Regrettably, Ladyman and Ross make only occasional reference to Sellars's images, as when they blithely remark how 'the locus of most metaphysical discussions is an image of the world that sits unhappily between the manifest image and an out of date scientific image' (2007: 10), and the theme of a regulative ideal of synoptic vision is never directly approached. The re-injection of such a synoptic ambition into philosophy of science, as I noted in the Introduction, is a most desirable and overdue goal.

Once a structuralist ontology is adopted, the issue remains: how to formulate a language allowing us to refer meaningfully to macroscopic objects while being ontologically committed to structures only? How to articulate a metaphysical worldview grounded on the primacy of physics and yet capable of legitimising not only everyday talk, but also the inquiries of the special sciences? To make sense of this position, seeking a unificationist but not naively reductionistic[55] solution which would avoid 'the faddish advertisements for "emergence"', Ladyman and Ross argue for the '*scale relativity of ontology*' (2007: 199) adopting the conceptual tool (following and revamping Dennett's ideas)[56] of 'real patterns'. Their structuralist metaphysics allows them to extend the common-sense/phenomenological insight that our strategies of description should vary according to the scale of the phenomena we are interested in, to the ontological hypothesis that 'claims about what (really, mind-independently) exists should be relativized to (real, mind-independent) scales at which nature is measurable' (2007: 200). The scale-relative pattern-based structuralism allows inquiry into 'regional ontologies' without demanding an underlying 'fundamental ontology'.

Patterns, most generally, are stable (described by law-like regularities) relations among a set of data.[57] Ladyman and Ross extract from Dennett's paper a necessary condition for the reality of a pattern: 'a "real" pattern, Dennett argues, must admit of capture using a smaller number of bits of information than the bit-map transcription of the data from which the pattern could be computed' (2007: 202), where 'could' is to be impersonally interpreted as a possibility, unrelated to any actual computation performed by a (human or non-human) computer. On the other hand 'the pattern cannot be compressed by any physically possible computer'

(2007: 206). A real pattern, then, is there to be identified in a data-set by *any* computer but it can be further compressed by *no* computer (under the constraints of the laws of physics). In both cases, there is an objective fact of the matter whether a real pattern *can* be discovered or *cannot* be compressed – hence the realism of this position. This picture puts ontological flesh on the structural bones of OSR: real, structural patterns replace objects at all scales. This move is eliminative insofar as it requires the pruning of objects from our metaphysics and it avoids – via the scale-relative location of patterns – the naïve reductionism of object-based ontologies. The two maxims '[t]o be is to be a real pattern' and "it"s real patterns all the way down' (2007: 233, 228) sum up the basic commitments of this ontology. The entities referred to in the talk of non-structuralist metaphysicians and laypeople alike are nothing but real patterns, structurally representable informational networks resisting (information-theoretic) entropic dissipation, capable of being tracked[58] across time, and, importantly, algorithmically *projectable* to the future. Only for epistemic convenience do we talk of individual objects (whether electrons, genes, species, or even persons): these are but 'epistemological book-keeping devices' (2007: 140) the ontological truth-makers of which are, actually, real patterns. From another perspective, though, it *does* make sense to talk about entities at all scales (from the chair on which I am sitting to the single electrons composing it) since all entities can be reinterpreted as *real* patterns at different scales of individuation – hence justifying the relative autonomy of all the special sciences (for Ladyman and Ross also including non-fundamental parts of physics), and of our manifest-image experience of things, thus solving Eddington's dilemma. As they put it: 'all real patterns except those of fundamental physics are detectable at some scales but not at others' (in Ladyman et al. 2013: 108). What is achieved is *unification* of the sciences *without reductionism*, free from notoriously unclear notions of 'emergence' or 'supervenience' to explain the descriptive efficacy of the special sciences and of our everyday conceptual frameworks.

Ladyman and Ross strongly emphasise the revisionary ambition of their position when they write that

> *The single most important idea we are promoting in this book is that to take the conventional philosophical model of an individual as being equivalent to the model of an existent mistakes practical convenience for metaphysical generalization.* We can understand what individuals are by reference to the properties of real patterns. Attempting to do the opposite – as in most historical (Western) metaphysical projects – produces profound

> confusion...Thus, saying that some patterns are real in a way that individuals are not is intended to express the idea that individuals are resolved out of patterns rather than vice versa. (2007: 229)

This is a strikingly ambitious programme, attempting to revise millennia of metaphysical speculation and undermine most contemporary work in the field. Their joint commitment to a renewed and naturalised verificationist criterion, restricting the range of meaningful assertions to those compatible with the empirically (physically) testable, importantly diverges from previous (logical positivist) attempts to disarm metaphysics through a rigorously applied condition of empirical verificability. To submit one's metaphysics to the unappealable primacy of physics does not restrain its explanatory ambition; on the contrary, it unleashes its true ambitions. So Ladyman and Ross argue that '[m]ost philosophical commentary on science that is rooted in conceptual analysis amounts, even if inadvertently, to efforts at domesticating science and ultimately restricting its reach' (in Ladyman et al. 2013: 112) and thus amounts to a philosophical conservatism which they 'abhor' and view as 'a sad refusal to explore the magnificent range of possibilities that our ability to do mathematics allows us' (ibid.: 113). As Russell memorably put it

> [o]rdinary language is totally unsuited for expressing what physics really asserts, since the words of everyday life are not sufficiently abstract. Only mathematics and mathematical logic can say as little as the physicist means to say. As soon as he translates his symbols into words, he inevitably says something much too concrete, and gives his readers a cheerful impression of something imaginable and intelligible, which is much more pleasant and everyday than what he is trying to convey. (1931: 85)

4.4 Mathematical structuralism

It is not my ambition to offer an unassailable defence of OSR and its ontology of patterns. I do believe it to be a well-warranted and convincing thesis, but all that is necessary, for my purposes, is to have shown that it is a *defensible* position, and indeed an active and promising research programme in contemporary philosophy of science.[59] The most far-reaching consequence of OSR, as far as my large-scale argument is concerned, lies in its offer of a naturalist, immanentist explanation of the post-seventeenth-century mathematisation of nature which has driven scientific effort from the Scientific Revolution[60] to the present, while side-stepping all Kantian cautions regarding the scope of our knowledge and the *a priori* nature of mathematical thought. Such a thesis – and this is my main argument here – can offer a much-needed naturalist supplementation to Badiou's philosophy, the metaphysical picture of OSR offering the best explanation for the efficacy of mathematics. At the same time, and this is but the other side of the same explanatory coin, Badiou's thought can offer meta-physical, or meta-mathematical, speculative insights building upon these naturalist grounds.

In the OSRist literature the possible identification of physical and mathematical structure is approached with overt caution. French and Ladyman acknowledge that 'the structural dissolution of physical objects leads to a blurring of the line between the mathematical and the physical' (2003a: 41)[61] but warn (replying to critics accusing them of unintelligible Pythagoreanism) that 'blurring does not imply identity. The mathematical can be trivially distinguished from the physical in that there is more of it; there is more mathematics than we know what to (physically) do with' (2003b: 75). Ladyman also voices his scepticism when observing that

> [a]s to what makes the difference between concretely instantiated mathematical structure (physical structure) and purely mathematical structure, I think that any attempt to say so would amount to empty words that would in the end add nothing to our understanding of the difference. I have no idea what conceptual resources one could deploy to say more about a distinction that, if it obtains, is so fundamental. (2009: 166–7)

Ladyman and Ross, however, are cautiously open to a possible speculative meeting between their OSR and structuralist positions in the philosophy of mathematics (converging towards the vanishing point of the obliteration of the abstract/concrete distinction), acknowledging that

> one distinct, and very interesting, possibility is that as we become truly used to thinking of the stuff of the physical universe as being patterns rather than little things, the traditional gulf between Platonistic realism about mathematics and naturalistic realism about physics will shrink or even vanish. The new wave of structuralism in the philosophy of mathematics, which has a number of supporting arguments in common with OSR…adds substance to this speculation. (2007: 236–7)

As they note, a plausible way of supporting the equivalence thesis is to suggest that the structuralist project in the philosophy of science can be supplemented by a parallel, but independent, trend in the philosophy of mathematics called, unsurprisingly, mathematical structuralism (henceforth, MS). Steven French (2014) has focused on precisely this link between OSR and MS while examining what he labels the 'collapse problem' – the idea that the structuralist position entails a collapse of the physical into the mathematical. While he resists the willy-nilly free exchange of arguments from MS to OSR, suggestion that it leads to confusions, his scepticism (and rejection of the aforementioned 'collapse') seems more motivated by the interest in defending OSR from the attacks of those who would use its similarities with MS as a way of critiquing the former's suitability as a realist stance (mostly, pressing on the issue of the causal inefficacy of abstract structures) than by a full-blooded rejection of the idea of a blurring of physical and mathematical. So, having noted that abstraction plays an ineliminable role in scientific modelling, he cautions against the equation of representational medium and represented reality noting that 'that the *representation* of physical structure involves such abstraction does not imply that the structure itself should be regarded as abstract, in the way that a mathematical structure might' (2014: 207). However, he goes on to admitting that '[t]his is not to say that there might not be reasons for holding such a claim [that the world has an abstract structure independent from concrete systems] and in the spirit of maintaining a 'Big Tent' approach, I'm prepared to entertain the possibility that some form of "structure as abstract" version of OSR is viable' (2014: 209). Ultimately, his stance on the issue remains an agnostic one: concluding his survey of possible ways of envisioning the relation between mathematics and physics, he admits (referring to the same texts I quoted above) that

> [p]erhaps then we simply have to accept that the distinction between the mathematical and the physical has, at the very least, become blurred (French and Ladyman 2003) or that it cannot be drawn at all (Ladyman and Ross 2007). Perhaps there is no answer to the 'problem of collapse'. Perhaps we should follow Ladyman in dismissing this as a pseudo-problem (2007, 2009). (French 2014, 230)

Some philosophers of mathematics have been far less cautious and non-committal in suggesting the equivalence of the mathematical and the physical,[62] so in order to explore the possible convergence between OSR and MS I will now survey the basic tenets of the latter, with the ultimate aim of highlighting the most promising option for a naturalised Badiouian ontology.

In recent decades mathematical structuralism has been emerging in a thoroughly elaborated form as a contender for the role of realist philosophy of mathematics. While varieties of structuralist positions vis-à-vis mathematics could be already found in Dedekind,[63] Hilbert,[64] Poincaré,[65] Benacerraf, Bourbaki[66] and Quine,[67] the approach has been proposed anew as the most promising philosophy of mathematics alongside the canonical positions of Platonism, formalism, logicism and intuitionism.

While a methodologically structuralist approach to mathematical practice is widespread in twentieth-century mathematics[68] the more articulated metaphysical formulation is due to Stuart Shapiro and Michael Resnik[69] in their respective monographs.[70] Analogously to OSR, MS intends to resolve the (Benacerrafian) problems related to the problematic existence of independent mathematical objects by reinterpreting them in structural terms, thus offering a robust realist stance as regards both ontology and truth-value. Just as the OSRist opposes the constructive empiricist's retreat into anti-realism when facing a metaphysically underdetermined physics, the MSist avoids the Benacerrafian rejection of ontological realism grounded on the set-theoretical underdetermination of natural numbers.[71] Like van Fraassen vis-à-vis scientific theories, Benacerraf accepts the structuralist approach to mathematics as recommending an ontologically non-committal approach towards mathematical entities.[72] MSists, on the other hand, stress how the basic conceptual units employed in the discussion of ontological commitment are mistaken and, like the OSRists, invite the re-conceptualisation of individual mathematical 'objects' as nodes in an ontologically primitive structure. Fraser MacBride (unwittingly) makes this parallel between the strategies of OSR and MS emerge clearly when he writes that

> perhaps it is not the mathematician's knowledge that is incomplete but the mathematical objects themselves: it is not ignorance that hinders the mathematician from settling whether 2 is $\{\{\varnothing\}\}$; it is the very absence of answers to such questions. It is to make sense of this idea – the idea that mathematical objects are incomplete in re – that some philosophers identify (or compare) these objects with positions in patterns or structures. (In Shapiro 2005: 564)

The mathematical structuralist holds that numbers have no independent existence over and above their embeddedness in an abstract structure, and that mathematics is the deductive study of structures (or patterns) as such. The slogan 'mathematical objects can be described only up to isomorphism' is true because there is nothing to know about them (they have no self-identity) except their structural roles.[73] Shapiro advocates an *ante rem* structuralism, where the structure or pattern itself is considered as a *sui generis* universal, independent of any particular instantiation of it (a *system*): numbers then are not to be interpreted as objects collectively composing a structure but, on the contrary, as places or positions within an ontologically prior structure:

> [e]ach mathematical object is a place in a particular structure. There is thus a certain priority in the status of mathematical objects. The structure is prior to the mathematical objects it contains, just as any organization is prior to the offices that constitute it. The natural-number structure is prior to 2, just as 'baseball defense' is prior to 'shortstop' and 'U.S. Government' is prior to 'vice president'. (1997: 78)

There are no ontologically primary objects needed to 'fill' the 'offices' (by analogy with a real-life organisation) of the structure: the distinction between office and office-holder simply disappears. Similarly, Resnik argues that

> in mathematics the primary subject-matter is not the individual mathematical objects but rather the structures in which they are arranged. The objects of mathematics, that is, the entities which our mathematical constants and quantifiers denote, are themselves atoms, structureless points, or positions in structures. And as such they have no identity or distinguishing features outside a structure. (1997: 201)

For the structuralist, for example, arithmetic is not the study of numbers *qua* objects, due to the latter's *incompleteness* (Resnik's term).[74] Rather, it is the study of the natural number structure, in whatever way that is exemplified in a system: '[t]here is no more to the natural-number structure than the relations embodied in the language and exemplified by the numerals' (Shapiro 1997: 138). Systems, the MSists claim, are instantiations of structures, collections of objects (not necessarily *physical* objects) and their relations modelling a certain abstract structure (in model-theoretical terms, 'system' can be rendered also by 'model' or 'interpretation'), so that 'structure is to structured as pattern is to patterned, as universal is to subsumed particular, as type is to token' (Shapiro 1997: 84).

A Benacerraf-style dilemma is thus defused: whatever relational system we examine (composed of either concrete objects or pure sets), the object of mathematics is the abstract, ontologically independent[75] structure that they instantiate (so two systems are isomorphic if there is a one-to-one correlation between their elements and those of their common structure – Resnik talks of '*pattern-congruence*' (1997: 204) when one or more concrete system(s) instantiate a pattern). Shapiro takes this as a form of Platonism (in the historical sense of a 'One-over-many' doctrine and not in that of the traditional object-based mathematical realism),[76] and calls it an '*ante rem* realism', opposed to the broadly Aristotelian view that the existence of abstract structures depends on the existence of their systems (*in re* realism). For the latter '[t]here is no more to the natural number structure than the systems of objects that exemplify it' (1997: 150).

A metaphysically deflationary variation upon the *in re* approach is often called eliminative structuralism,[77] and it rejects the existence of both objects and structures *qua* universals altogether. The eliminativist agrees with the *ante rem* structuralist about the conceptual priority of structures over systems: statements about the latter *imply* a reference to structures. But, the eliminativist continues, they do not *entail* that such structures exist and enjoy ontological priority: they should simply be seen as generalisations, convenient verbal shortcuts without metaphysical consequence. Keeping to the 'places' and 'offices' analogy, Shapiro (1997: 82, 83) notes that this eliminative approach prefer a '*places-are-offices*' approach to the '*places-are-objects*' one of the *ante rem* structuralist. The basic ontological difference between the two is that while the latter reconceptualises 'objects' as places, and hence takes the structure comprehending those places as ontologically primary, the former requires (*if* realism is a desideratum) some sort of background ontology: if places are offices, *something* must occupy them. As Shapiro explains 'in the case of chess games, the background ontology is small, movable objects – pieces with certain colors and shapes. In the case of arithmetic, sets – or anything else – will do for the background ontology' (1997: 82). On the other hand, as we have seen, the *ante-rem* structuralist adopting a places-are-objects perspective will be happy to claim that '[b]ona fide singular terms, like "vice president," "shortstop," and "2" denote bona fide objects' (ibid.).

4.5 Structures *of* what?

I can now start to retrieve some loose ends of my large-scale argument by going back to Maddy's set-theoretic realism expounded in Chapter 2. Maddy sees a substantial agreement, both ontological and epistemological,[78] between her position and the one of the MSist, but notes that

> [w]here the two part company...is in their view of the set theoretic universe itself. For the set theoretic realist, this 'structure' consists of real objects, the sets: these are the bedrock, the things that instantiate the various mathematical universals. For the structuralist, it is just one more structure, made up of featureless points in certain relations. (1990: 173)

However, she goes on to notice how a fully independent structuralism would need to offer a complete axiomatised theory of structures or patterns, able to tell us which structures exist. This is precisely what Shapiro does (1997: Ch. 3), listing a series of axioms which ends up being essentially equivalent to that of set theory. He indeed remarks that '[t]alk of structures, as primitive, is easily "translated" as talk of isomorphism or equivalence types over a universe of (primitive) sets. In the final analysis, it does not really matter where we start', and considering that 'on any structuralist program, *some* background theory is needed' (1997: 96) the choice between *ante rem* structure theory and set theory is purely a matter of taste: 'they all say the same thing, using different primitives' (1997: 97).

At this juncture, then, the most tempting option for my purposes is to adopt what Shapiro defines as an ontological eliminative structuralism, a structuralist position denying the existence of *ante rem* abstract universals (Shapiro's structures) and offering set theory as a background ontology for all mathematical structures: this position denies independent existence to both mathematical objects (over and above their structural role) and *ante rem* mathematical structures and adopts a places-are-offices perspective. As noted above, any such position needs a rich enough background ontology, provided by the set-theoretical universe.

Note, however, that this requires me to return to Benacerraf's dilemma[79] and reassess my comparison with the fork between constructive empiricism and OSR. While it is true that the *ante rem* structuralist response to Benacerraf is similar in form to the OSRist alternative to van Fraassen, a crucial metaphysical difference obtains. The OSRist claim that 'there are no objects, just structure' can be made intelligible to the extent that our ontology, having lost a mooring in objects, can 'fall back' on mathematical structure (a move, as we have seen, justified by a commitment to the methods of the mathematised natural sciences). However, the same claim *cannot* be made intelligible by MS without postulating a *new* kind of entity (the abstract, universal, *ante rem* structure). In the mathematical register the MSist who denies the existence of any object has, so to speak, nothing to fall back onto, and is forced to postulate an ad hoc safety net. This is a move that the naturalist wants to avoid, the immanentist aim being the articulation of a *continuous* worldview, ranging from the mathematical to the physical without ontological breaks or gaps between different domains.

It seems to me that this 'Aristotelian' move allows us to discern the mistake in a view, such as Shapiro's, that attempts to explain the applicability of mathematics to physics through reference to universals. I thoroughly agree with Shapiro's contention that '[o]ne cannot begin to understand how science contributes to knowledge without some grasp of what mathematical/scientific activity has to do with the reality of which science contributes knowledge' (1997: 244), and that therefore to offer an account of the relation between the mathematical and the physical is an unavoidable task for the epistemologically aware philosopher *qua* philosopher (not a task restricted to philosophers of mathematics). On the other hand, his proposed resolution to the problem, invoking a new class of universals, closes an explanatory gap by opening a metaphysical one. Shapiro writes that

> the relationship between mathematics and material reality is, in part, a special case of the ancient problem of the instantiation of universals. Mathematics is to reality as universal is to instantiated particular... More specifically, mathematics is to reality as pattern is to patterned... A first attempt to articulate the present thesis would be that science proceeds by discovering exemplifications of mathematical structures among observable physical objects. (1997: 248)

Along the same lines, Dorato writes that

> the simpler explanation for the applicability of mathematics... consists in holding that the physical world exemplifies abstract mathematical structure as a matter of fact, in the same sense in which three mountain tops and three marks of chalk on a blackboard are concrete instantiation of the abstract structure of a triangle. (In Antonini and Altamore 2010: 68)

As many have shown, it is impossible (or it is possible only by neglecting the actual practice of applied mathematics) to claim that, for example, the regular motion of a pendulum, the vibration of a string or the orbital periods of the moons of Jupiter are a concrete instantiation of the universal mathematical structure describing their behaviour, for only in highly idealised circumstances does such an isomorphism obtain.[80] Unlike Shapiro I do not think that a generalised 'observation that the contents of the nonmathematical universe exhibit underlying mathematical structures in their interrelations and interactions' (1997: 248) is a solid enough evidential basis to reach his desired conclusion. Let me stress again that I want to argue for the case that it is *at an elementary scale*,[81] where concrete reality and mathematical structure blur into one another; I am neither interested in 'mathematising away' Jupiter and its moons nor in postulating the existence of some hyper-platonic, *sui generis* mathematical structure 'out there' (*where*? This is the deal-breaker for the naturalist) which the Jupiter-and-moons system happens to instantiate. It seems patently absurd to claim that there is nothing more to Jupiter and its moons than the relations expressed in the (mathematical) gravitational laws we use to describe and predict their movement. It is far less contentious, for example, to offer a group-theoretical account of 'objects' according to which unobservable *elementary particles* can be exhaustively (and objectively) described as irreducible representations of fundamental symmetry groups[82] – take this as a *restricted*, or *revised*, form of the indispensability argument.

A compromise position between a full-scale structuralism (claiming that all objects depend on abstract, *sui generis* structures) and a non-structuralist Platonist stance (claiming that no mathematical object is so dependent) is defended by Øystein Linnebo. He argues that the structuralist thesis of ontological dependence (on other objects or on the entirety of structure) holds true for the majority of mathematics, *except* sets. He explains that

> sets provide examples of mathematical objects that are not subject to the upwards dependence ... According to the prevailing iterative conception, sets are 'formed from' their elements. The relation between a set and its elements is thus asymmetric, because the elements must be 'available' before the set can be formed, whereas the set need not be, and indeed cannot be, 'available' before its elements are formed. A set thus appears to depend on its elements in a way in which the elements do not depend on the set. (2008: 72)

Moreover, Linnebo continues, this allows us to think of the set-theoretic hierarchy in a piecemeal fashion, so that 'we can give an exhaustive account of the identity of the empty set and its singleton without even mentioning infinite sets' (2008: 73). In sum, then, the iterative hierarchy of the set theoretic universe, grounded solely on the existence of the empty set is enough of a background ontology for a structuralist approach to mathematics that wants to avoid the reference to abstract universals. Mathematics is a science of structures (that is, indifferent to the nature of the entities composing particular structures), and structures are composed of 'featureless' empty sets and their relations. Our ontological commitment can be reduced to one kind of entities: pure sets.[83] The adoption of an ontology of pure sets is not at odds with the structuralist approach, since the latter's central idea is that 'mathematical objects do not have a richer "nature" than is given by the basic relations of some structure in which they reside' (Parsons 2008: xi).

Given that an ontology of universals, the naturalist argues, should be rejected,[84] this kind of compromise position, restricting our ontological commitment to pure sets (such that only conventionally, or in manifest-image talk, can we meaningfully talk about impure sets at all), allows us to construct mathematico-physical structures without reintroducing irreducible and unexplainable dualisms (between universals and particulars or between concrete and abstract) into our metaphysical worldview.

Having described, in this chapter, the possible encounter between the independently defensible structuralist tendencies in the philosophy of science and the philosophy of mathematics, in the next chapter I will argue that a Badiou-inspired approach might, in turn, prove to be a germane speculative supplement to this structuralist, matherialist metaphysics.

5
Truth and Randomness

In the preceding chapter I have attempted to keep the Badiouian references to a minimum, so that the material presented therein could stand on its own merits without being subjected to Badiou's authority or veto. The synthesis I strive for between Badiou's ontology and the metaphysics of structure in this chapter is not guided by Badiou, but by my own synoptic commitments to a philosophy making free (but informed) use of heterogeneous sources. The aim of this final chapter, then, is to delineate further and more concretely than I have hitherto done a mutual supplementation of Badiouian ontology and structural realism by suggesting how the former's stance towards truth and ontological incompleteness can be naturalistically reinterpreted as compatible with the latter

5.1 Against fundamental ontology

As I explained in Chapter 2, I believe that the Badiouian insistence on a project of fundamental ontology should be abandoned as a post-Heideggerian morass, irremediably tainted by its reference to a 'Being' beyond presentation (or below, or before, or retroactively individuated/subtracted from: no amount of spatial, temporal or arithmetical metaphors adequately clarifies this concept), that even Badiou's 'subtractive' approach does not fully manage to demystify.[1] As Johnston argues, 'one of several combative slogans for any contemporary materialism... is "Forget Heidegger!"' (2013: 93). Badiou's regular subordination of 'Nature' to 'Being' (arguably explaining his unfamiliarity with the results of the physical sciences) is an evident symptom of this betrayal of immanentist principles, and engenders the unresolved[2] quandary of the relationship between the empirical and the ontological. Commenting on Galileo's 'very drastic affirmation' (the identification of the language of nature with mathematical language) Badiou explains:

> I think it is not true, in fact. It is not exactly Nature which is written in mathematical language. Because Being is not reducible to Nature. Nature is something like Being with the idea of the World. So Being and localisation in a World, that is Nature... in general. But for Galileo Nature was a name for Being as such, Nature was the Being of beings or something like that, and so Nature was written in mathematical language. (2012: n.p.)

And later adds that

> in Heideggerian terms, in some sense, we can say something like that: we have no knowledge of Being, we have only thinking of Being. We have no empirical knowledge of Being as such, we have only thinking. But... what is reducible to thinking as such is only mathematics. All what [sic] is not of mathematical Nature is empirical in some sense. (2012: n.p.)

These are distinctly telling pronouncements. The naturalist philosopher, for whom the term 'nature' indexes *all there is* (in the broadest possible sense of the term), cannot accept this Badiouian rift between a knowable 'nature' and a thinkable but unknowable 'Being'. To be sure, Badiou's prioritisation of mathematics as a means for the active thinking of Being is preferable to Heidegger's predilection for its passive, poetically articulated reception (to the extent that the validity of mathematical inferences can be tracked and rationally assessed, utterly unlike poetic *Erhabenheit* which, rejecting discursive articulation, leans dangerously toward undiscerning and unaccountable authoritarianism).[3] Moreover, Badiou's insistence on the projected *infinity* of the rational subject is a welcome distancing from the rhetoric of finitude that characterised Heideggerian and post-Heideggerian thought. However, to invoke a 'Being' beyond any *knowledge* to warrant, as in his doctrine of the event, the possibility of new *truth* is a metaphysical (if not downright *theological*) hang-up we should relentlessly seek to discard. Is there not a certain convergence between Heidegger's denunciations of techno-science as the contemporary historical mode of forgetting of Being and Badiou's distinction between the Being presented by mathematics and the mere 'Nature' studied by the rest of the empirical sciences?[4] Only at the price of reducing all science to abstract mathematics can Badiou distance himself from Heidegger's infamous claim that '[s]cience does not think' (1993: 373), that is does not think the Being of beings. Both Badiou's and Heidegger's overwhelming concern with giving a new formulation to the question of Being systematically devalorises epistemic inquiry and, in general, the importance of epistemological questions.

We can certainly agree, with Badiou, that the *Seinsfrage*, once poetic pathos is replaced with mathematical indifference, rings hollow of meaning – this is one of Badiou's most valuable insights: the privileging of mathematics as a discourse unconstrained by the lures of meaning where '[i]n mathematics, the Real is shown to be deprived of sense' (2006c: 56). But we should also push the demystification further and reject the notion that mathematical presentation needs anything more than itself (such as a non-discursively accessible real 'Being') and indeed question whether it is a 'presentation' of *anything*. The naturalist, in sum, is unconvinced by Badiou's half-hearted replacement of poetry with mathematics. His mobilisation of the latter, *mutatis mutandis* (the main difference being Badiou's insistence on the *multiplicity* of Being, which makes it impossible for it to be captured by a single concept), preserves intact the structure of the Heideggerian problem of a difference between Being and beings,[5] or, at worst, spawns a perplexing *triad* of (1) inconsistent multiplicity, (2) consistent/presented multiples and (3) natural multiples/physical reality. Badiou himself explicitly suggests that his Number and numbers distinction should be interpreted as another formulation of the ontological difference,[6] explaining that

> [w]e might also say that between Number, which inscribes its section in the unrepresentable inconsistency of natural multiples, and number, which we manipulate according to structural links, passes the difference between Being and beings. Number is the place of the being qua being, for the manipulable numericality of numbers. Number ek-sists in number as the latency of its being. (2008a: 211)

Having avoided the 'Great Temptation' of negative theology is Badiou not falling prey to the equally pernicious temptation of expecting something, even an *inconsistent* something, under the differential surface of mathematical structure? Whilst Badiou's concern with the discursive intelligibility of Being as always mediated by its mathematical presentation allows for a rational, disenchanted treatment of Being which is certainly more exhaustive than Heideggerian mystical allusiveness (if at the price of rendering the *thought* of Being and Being *itself* one and the same), the chasm he opens between thought/Being and nature becomes either unbridgeable or inexistent. As Brassier puts it:

> Badiou finds himself confronted by two equally unappetizing alternatives. On the one hand, he faces a relapse into the empiricist dualism of formal scheme and material content, which he himself had previously sworn to abjure... Or, on the other hand, there is the prospect of a discursive variety of absolute idealism – or crypto-Hegelianism – in which the difference between the conceptual and the extra-conceptual, or discourse and world, is reduced to the distinction between consistent and inconsistent multiplicity, and for which, ultimately, thinking is all that matters. (2007: 115)

Badiou's thesis claims that

> two directions, two orientations command the entire destiny of thought in the West. One, based on nature in its original Greek sense, welcomes – in poetry – appearing as the coming-to-presence of being. The other, based on the Idea in its Platonic sense, submits the lack, the subtraction of all presence, to the matheme, and thus disjoins being from appearing, essence from existence. (2006a: 125)

But the naturalist remains unconvinced that we should take either the poetic or the subtractive route, for the reason that both lead to a problematic reference to a 'Being' which can be only poetically named or mathematically thought, but never *known*.[7] If the Badiouian–Platonic stance 'disjoins being from appearing' it does so only at the price, as I noted in Chapter 2, of conflating Being with thought itself, thus locating the real squarely on the side of the intelligible.

A structuralist ontology allows us to deflate the question: talk of Being as a condition of possibility, horizon, framework, or domain for beings to appear in (or for situations to be counted-as-one out of) can be fully replaced by the talk of a structure as the framework for 'objects' to appear in. To 'be' is to be structurally individuated. The significant difference, of course, is that 'structure' can be given a mathematical definition, demystifying a 'Being' which *defies* definition (at least as anything more explicit than 'inconsistent multiplicity'). Against Badiou, then, my claim is that, *yes*, 'Being' is reducible to 'Nature', since 'Nature' is (mathematically articulated) structure all the way down. *There is nothing outside the structure.*

I am here explicitly rejecting the distinction, articulated by Dale Jacquette, between a pure philosophical ontology and an applied scientific ontology, or at least invalidating any inquiry into the former. Jacquette (along lines already drawn by Husserl, with the distinction between *formal* and *material* ontology) explains that

> [p]ure philosophical ontology is concerned with the meaning of the concept of being, with the question why there is something rather than nothing, and the modal ontological status of the actual world. Applied scientific ontology builds on the conceptual analysis of what it means for something to exist in order to recommend a preferred existence domain, thereby committing itself to the existence of a particular choice of entities. (2002: 4)

It follows from this demarcation that no applied ontology can begin before a pure ontology has been established. Interestingly, Jacquette opens his discussion of pure philosophical ontology with a critical examination of Heidegger's project of fundamental ontology, which he rejects, like Badiou, on account of its methodological shortcomings (not the soundness of the enterprise). Jacquette argues that the (early) Heideggerian conflation of phenomenology and ontology is an unjustified premise, adopted by fiat under the assumption that a systematic phenomenological reflection on the human-centred ways of givenness of the world can lead to the articulation of universally valid ontological insights. 'We cannot', Jacquette argues, 'be limited in scientific metaphysics to the anthropocentrism entailed by ontology as phenomenology' (2002: 37).[8] Again like Badiou, Jacquette seeks another foundation for pure ontology, one fully disentangled from anthropocentric bias (that of Heidegger's existential analytic of *Dasein*), and finds it (unlike Badiou) in logic. What needs highlighting here is that Badiou's and Jacquette's agreeable rejection of Heideggerian phenomenologically rooted fundamental ontology are both likewise accompanied (if not motivated) by a disdain of empirical investigation *tout court*. As Jacquette most explicitly puts it in his critique of Heidegger: '[p]ure philosophical, as opposed to applied scientific ontology...cannot rely on the empirical data of sense experience, no matter how impressive, vivid or compelling' (2002: 36). While I am happy to preserve some privileged role for *a priori* inquiry in philosophy – and I certainly do not endorse the naive empiricism Jacquette seems to adumbrate – I am not persuaded that the high road of fundamental ontology, utterly detached from empiricist scruples, can lead to any philosophically interesting place. Likewise, I am not sure that the time-honoured question 'why is there something rather than nothing?' *can* be given an answer (at least in this too general a formulation). Should this seem like a rather quick dismissal of an ontological project that is one of the backbones of the Western philosophical tradition,[9] I cannot but reply that such fundamental disagreements set apart incommensurable philosophical stances: to those arguing that an ontology of scientifically describable entities cannot be defensible without a prior (and *a priori*) clarification of what 'to exist' *really* means I feel compelled to object that since at least Kant (and surely since Quine's predication-based notion of ontological commitment) we dispose of different possible strategies to deflate this seemingly profound question.[10] More cogently still, I believe that the adoption of a structuralist worldview implicitly undermines the Aristotelian question of Being *qua* Being, to the extent that the latter was formulated through the explicit priority Aristotle assigned to the category of substance (*ousia*) as the underlying substrate (*hypokeimenon*) to which all accidental attributes would inhere, precisely the substantialist paradigm that the structural realist claims our best current science impels us to reject.[11] Thus I would say with Aristotle himself 'if there is no substance other than those which are formed by nature, natural science will be the first science' (*Meta.* E.1, 1026a 27–9). That is right: there isn't and, I should add, if there is *no substance* at all, mathematics will be first science.[12]

Does this mean that I place myself squarely within the 'analytic' approach to ontology? Early in their recent introductory textbook to metaphysics, Carroll and Markosian explain to their readers that

> [s]ince at least the mid-twentieth century, ontologists have not primarily studied or tried to characterize the concept of existence, at least not directly. It would be more accurate to describe them as trying to complete the does-exist list. They have not primarily addressed the question, 'What is it to exist?'; they have primarily addressed the question 'What exists?' So do not think of ontology as starting with some pre-theoretic judgements about what exists and what does not, and then trying to formulate a description of the crucial differences. Better to think of ontologists as doing inventory. (2010: 12)

If this indeed is an accurate description of analytic ontology, I significantly diverge from it in a crucial, Badiouian, respect: the philosophical project cannot be limited to a humdrum enumeration of existent entities or kinds of existing entities[13] and I believe that this lack of ontological ambition is a symptom of (in Badiouian parlance) the timid 'constructivist' approach buttressing this stance. Ontology should have higher aspirations than merely listing what we quantify over in our best scientific theories. We can and should aim at breaking the confines of the encyclopaedia of knowledge towards novel truths, to cultivate a creative metaphysics capable of making 'a predicate of the impredicable' (Badiou 2000b: 183). Sometimes to follow scrupulously a principle of least epistemic risk[14] is ill-advised, and to relax slightly that constraint for ontological speculation could pay philosophical dividends. Indeed (as I noted in closing in the previous chapter) I believe that the orthodox naturalism and scientific metaphysics guiding the work of structural realists can and should be enriched by *meta-ontological*, rationalist reflections of Badiouian inspiration, maintaining the reality of novelty and indeterminacy. So, in agreement with Paul Livingston, I too believe that

> an ascendant empiricist 'naturalism' increasingly precludes thematic access to the profound historical results of rationalist thinking about logic and metalogic that once defined the greatest achievements of this tradition itself, and remain some of the most important accomplishments of twentieth-century thought. (2011: xiii)

I therefore demonstrate the possibility of (1) exploiting the Badiouian meta-mathematical and anti-theological reflections while (2) defending a structural realist position capable of locating all existents on a continuum of structural complexity (beginning with the *Ur*-element of the empty set).[15]

This will lead to a synthesis capable of demonstrating how meta-mathematical and meta-logical structural paradoxes of self-reference and undecidability, of totality and transcendence of totality (inhering in *any* structural system and conditioning our knowledge of them) are to be seen as features of *reality itself*, not merely of our thought-patterns or of 'abstract' formal languages.[16] If there is nothing more to physical reality, at the elementary level, than mathematically describable structures, then *meta-mathematics is meta-physics* (or vice versa, the distinction is moot). Moreover, Badiou's anti-constructivist reflections on the immanent production of 'generic' truth out of current knowledge can, when reintegrated within a naturalist ontological picture, offer epistemological insights into the boundary-transcending power of reason, capable of exceeding its own conceptual limits towards a better grasp of reality, yet not productive of novelty itself. As Fraser (to my knowledge the only interpreter who has suggested a link between Badiouian ontology and OSR) has perceptively seen, the problem is that

> [i]n so far as the generic, by definition, exceeds algorithmic projectibility, there can be no generic 'real patterns', and so no 'real' 'truths' ('real' in OSR's sense, 'truths' in Badiou's). Philosophical naturalism finds its zenith in the radical conformity that OSR enforces between being and knowledge, and in doing so finds the greatest point of discord between its project and Badiou's – a discord that, perhaps, does much to explain the distance Badiou has maintained from naturalisms of every stripe. The Badiouean's task would be to carve a path for an unnaturalizable, but non-supernatural, subjectivity in the mathematics of real-patternhood. (In Bartlett and Clemens 2010: 182)

To suggest that such a discord can be resolved, to submit that we need not be bound to (what I will denounce as) Badiou's voluntaristic and anthropocentric conception of truth procedures (which, in my matherialist context, would end up suggesting a 'participatory universe')[17] and to attempt to lay down the contours of a synthetic encounter of these positions, is the aim of this chapter.

5.2 Idealism and rationality

The first task to undertake in order for my project to succeed is that of dispelling the looming charge of idealism. How can we defend the equivalence of reality with its structural/mathematical description without falling into mathematical idealism, absolutising our mathematical thought to the structure of the in-itself or postulating a pre-established harmony between real matherial structure and the conceptual structure of thought? But also, as I noted in Chapter 4 when surveying critiques of OSR, the question 'why should there be no gap between epistemology and ontology?' needs an answer. Does the rejection of the neo-Kantian stance of ESR necessarily force upon us the Hegelian conclusion that there can only *be* things we can *know* about (or in OSRist terms, that there is no-thing where nothing can be known of it), equating the structurally real with the relationally rational, and collapsing epistemology into ontology? These are crucial questions, since what is ultimately at stake is the very notion of a rational order of nature which I flagged, in Chapter 3, as *the* presupposition a naturalist philosopher of science should be relentlessly critical of. They are also tied to the problem already approached in Chapter 2: what ontological outlook can reconcile (1) a bottom-up inquiry into biological, evolution-driven abilities of mathematical conceptualisation (thus rejecting talk of mathematical axioms *qua* laws of thought or synthetic *a priori*)[18] and (2) the top-down OSRist thesis that physics unequivocally suggests a mathematical-structural universe? Can we endorse both the claim that mathematical thought is a hard-won cognitive achievement *and* the claim that contemporary science has revealed to us that there is nothing more fundamental to be offered in the way of a description of reality at its elementary scale than a mathematical structure?

The resolution lies in the overcoming of the binary of empirical content and abstract form, an idea I introduced in the previous chapter and one relentlessly pursued by the *early* Badiou in his materialist epistemology. While I share Badiou's hostility to the 'ideological' reduction of the mathematical to mere formal tool, I part ways with him when he chooses to collapse the distinction *on the side of thought,* offering autonomy to purely rational mathematical scriptural production and ontologising the intelligible (reading the equation thought = Being, so to speak, from left to right). There is nothing more to the matter/form distinction than there is to the abstract/concrete one, but rational thought is ontogenetically possible thanks to the pre-noetic existence of objective and extra-mental structure: the real is the causal antecedent of the conceptual. Putting the conceptual and the real on the same ontological continuum should not imply either panpsychism (everything thinks) nor idealism (thought is everything). Badiou's error here consists in the conflation of *formal* with *rational.* Reason is a *product* of pre-rational matherial reality, not an attribute of formal structure. For idealism to be rejected, rational thought must be *different* from what it thinks, since it thinks what it thinks only by virtue of a conceptually mediated sensible encounter with it; yet being necessarily *part* of it. A demystified understanding of rationality *qua* the competent, rule-bound articulation and application of concepts and their inferential relationships by sapient subjects may help us see how 'reason' cannot be ascribed as a feature of a non-cognitive agent, but rather is the capacity to proficiently engage in discursive practices. As Brandom puts it: '[t]he Enlightenment disenchantment of the world and its assignment to us of responsibility for the norms, values, and significance we nonetheless find in the world are two sides of one coin. Meaningless objects and meaning-generating subjects are two aspects of one picture' (1994: 49). Meaning depends on norms, and rules exist only as practical commitments within a society of rational agents, not as transcendent, divine, commands.

This problem can be approached through a discussion of the Principle of Sufficient Reason's (PSR) applicability. For Leibniz, the philosopher responsible for the principle's name, the PSR states 'that nothing happens without a reason why it should be so rather than otherwise' (1989: 321). For Spinoza, on the other hand, the principle states that 'for each thing there must be assigned a cause or reason, both for its existence and for its nonexistence' (2002: 222; 1p11d2). Leibniz's more traditional theism requires us to interpret 'happens' as 'it is so disposed by God',[19] so that the principle in effect supports the ontologically strong claim that God exists and does not act without a reason but always in accordance with maximal goodness, '*God does nothing which is unorderly*' (1989: 306). Spinoza, less interested in ensuring the *best* ontological picture than he is in describing the *truest* one, assigns a more radical role to the PSR, one where 'reason' is not much a top-down efficient cause but a possible step in a bottom-up inferential thought-process. Spinoza is not so much concerned with ideal intelligibility by a divine intellect but rather with *complete intelligibility*, that is the demand that everything that is can, in principle, immanently be given a reason by intellects like ours. Each fact, everything that exists, has an explanation. More than that, as Michael Della Rocca explains, Spinoza is thoroughly committed to a *naturalist* unification of explanatory principles: Spinoza's naturalism is the thesis that 'everything in the world plays by the same rules... [and] is the view that there are no illegitimate bifurcations in reality' (Della Rocca 2008: 6). Nothing is in principle inaccessible to human reason (there are no brute facts), as the latter is capable, at least in principle, of following the infinite chain of causes since its explanatory parameters are the very same ones at play in the universe at large. Where we can (and should) supplement Spinoza is in the understanding of 'infinite' in the previous sentence: exploiting the mathematical notion of limit[20] (unavailable to Spinoza), we can see that the procedure of understanding, by the finite mind, of an infinite series of causes can be seen to asymptotically approach but *never* reach its result: the terminus of the explanatory inquiry (or, better, of a convergence of several historically deployed projects of inquiry)[21] is determinate (*not* ineffable) yet only as an infinitely remote, receding limit. No *parousia* of rational intelligibility is ever to be reached in actuality (nor is it God's task to ontologically bolster actualised intelligibility: *Deus sive Natura* is the infinite process of infinite intelligibility). Rational intelligibility is always diachronic. So the Spinozist rejection of metaphysical bifurcations, endorsed by the naturalist, should not make it impossible to prise apart a realm of reasons from one of causes. That is to say, *reasons* are procedurally given in history by rational agents operating within an inferentially articulated nexus of concepts, whilst the realm of natural *causes* exceeds (both temporally and logically) that of reasons and, in its exorbitance, cannot be meaningfully assessed as intrinsically reason-governed. As Brandom vividly puts it:

> [o]ur activity *institutes* norms, *imposes* normative significances on a natural world that is intrinsically without significance for the guidance and assessment of actions. A normative significance is imposed on a non-normative world, like a cloak thrown over its nakedness, by agents forming preferences, issuing orders, praising and blaming, esteeming and assessing. (1994: 48)

So the *formal* nature of mind-independent reality can be explicated in terms of an intrinsic lack of any substantial or semantic content. Moreover, the possibility of algorithmically ascending the ladder of reasons does not by itself establish anything about the ontological completeness or consistency of the examined reality as a whole.

As for the charge of ontologising the intelligible: rejecting Badiou's Platonic ontological commensurability of known and knower (Being and thought), the matherialist holds that only through empirical investigation do we come to the conclusion that the universe is structural, an inquiry conditioned by our cognitive make-up and developed within our battery of inferentially articulated concepts. Indeed, the most important insight shielding structural matherialism from idealism is that our mathematical abilities (both the conceptualisation of structures and their deductive and abductive inferential connections) are bounded by biological constraints. This cognitive closure entails that there are parts of the structure of the universe that we probably cannot presently grasp. However, we are not stuck in a closed circle of ignorance: abductive methods, by bootstrapping our conceptual framework beyond its limits, continuously increase our knowledge. And yet, it would be a betrayal of Cantor's insights to assume that even an infinite progression of knowledge can ever fully map the transfinite realm of matherial reality. Parts of it will always exceed our grasp, *without for this reason being construed as transcendent*, nor, in Wittgensteinian fashion, as a non-entity about which we can only ever obtain an inexpressible insight.[22]

This is one of the advantages of the matherialist thesis: we as finite cognitive agents are neither endowed with omniscient rational insight nor is our epistemic grasp curtailed by some universe-transcending power. Both what we can and cannot know can be *immanently determined*. This is how the closure of principled gaps between epistemology and ontology is brought about: there is certainly more structure than we can presently or (perhaps) ever come to grasp, but this does not erode our certainty that there is nothing more (whether transcendent or noumenal entities) than structure. Structure is all there *is*, yet we cannot in actuality know *all* there is.

Exploiting the familiar Chomskian distinction,[23] the matherialist thesis entails the rejection of *mysteries*, whilst admitting *problems*. The latter need not be issues that can be solved by creatures like us at our *present* cognitive-evolutionary and intellectual juncture, but simply ones that do not lie beyond any principled boundary. We might lack the conceptual resources (or, as I will put it later, the computational power) to reach a certain truth about the universe, but this does not make it *mysterious*. How can we make sense of the notion of mystery (an intrinsically unanswerable question) when (1) accepting an evolutionary worldview which allows for the endless adaptive development of our cognitive powers, (2) considering the very real possibility of a technological supplementation of our brain (a post-human future) and (3) defending a computational understanding of rational processes and which, while excluding the possibility of reaching a fixed answer, nonetheless guarantees an infinite process of clarification? If the aforementioned mystery is not endogenously determined by our neurobiological make-up (for any such constraint cannot be proven unsurpassable, envisioning our cognitive capcities enchanced via biological or technological means) nor established by an omnipotent God capriciously delimiting the domain of the intelligible, what would guarantee its insolubility?

Borrowing Kant's useful terms, the rational progress of knowledge can be said to have *limits* [*Schranken*] as 'mere negations that affect a magnitude insofar as it does not possess absolute completeness', but no *boundaries* [*Grenzen*] which 'always presuppose a space that is found outside a certain fixed location, and that encloses that location' (Kant 2004: 103–4; §57, 4:352). We need not give assent to either the rationalist lures of omniscience nor the inauthentic admonitions of those preaching, with sham humility, a heteronomously imposed unknowability: the self-correcting and progressive power of rational thought can be defended without having to sit on the idealist's quaint intellectual barricade. We are biological creatures instantiating rational thought patterns, and the contingent limits of the former can be overcome by allegiance to the unconstrained potential progress of the latter.

Such is the outlook offered by a cognitive naturalism of pragmatist pedigree running from Peirce to Sellars: a thoroughly post-Darwinian paradigm where knower and known are placed on the same immanent plane, and where rational practices and physical processes are seen as part of the same natural continuum, subject to the same constraints and both undergoing evolution.[24] This paradigm, defending empiricism through a notion of experience as holistic process involving both subject and object, circumvents the pitfalls of what Badiou denounced in his early work as 'bourgeoisie epistemology' (post-Cartesian representationalism, pitting subject and object against each other in an epistemological standoff plagued by the eternal problem of faithful *mimesis*), while at the same time avoiding Badiou's rationalist excesses (and his undiscerning anti-representationalism) leading him to such proclamations as:

> philosophy does not dedicate itself to the care of the limits, but to the care of the unlimited...Not only, and contrary to what Hamlet declares, is there nothing in the world which exceeds our philosophical capacity, but there is nothing in our philosophical capacity which could not come to be in the reality of the world. It is this coextensivity *in actu* of conceptual invention and of a reality-effect that is called the absolute, and it is this that is the sole stake of philosophy. (2000b: 185, 189)

So, against Badiou, I hold that there conceivably are parts of reality that remain beyond what is rationally accessible to us humans, without for this reason being a-rational and *in principle* inaccessible ones. Badiou's explicit projection of the conceptual onto the real is disqualified: to assume that the range of the real cannot exceed the range of the knowable is an unambiguous mark of idealism. The Kantian project cannot be so easily dismissed as a conservative policing of boundaries, but should instead be reactivated, in a suitably modern form, as a propaedeutic for the accountable overcoming of boundaries. This would mean rejecting the most problematic facets of Kantianism which, ironically, linger in Badiou's very distinction between inconsistent and consistent multiplicities and in his dual register of ontology (in *Being and Event*) and phenomenology (in *Logics of Worlds*).[25] It would amount to a responsible attempt to draw the outline of our cognitive limitations without postulating ontological discontinuities: an unavoidable first step for those who aim at, through careful conceptual engineering (and arguably with the aid of computational technologies augmenting our own cognitive resources), transgressing those very limits.[26]

5.3 Truth as abductive process

As already noted, my two-fold move amounts to a double naturalisation of Badiou's ontology: on the one hand, the neurophysiological and cognitive-historical explanation of the genesis and development of mathematical concepts is consonant with a programme of naturalised epistemology; on the other, the structural realist identification of the physical with the mathematical is an IBE-justified naturalist move which reinscribes mathematics among the empirically validated sciences. Our mathematical thought, then, neither 'creates' nor fully and synchronically 'contains' the whole of reality, in its current or projectable state. Here is where the importance of the paradoxes revealed by metamathematics emerges most clearly, and where Badiou's conception of a 'generic' truth bears upon the structuralist position, since the latter can be endorsed without the 'constructivist' (in Badiou's terms) belief in a full capture of the real by the (current) concept, or complete equivalence between what is and what is knowable. A structural universe can produce real and unforeseeable configurations of itself: indeterminacy and contingency can be an object of fully rational, mathematical assessment.

The primary aim of Badiou's *Being and Event* is the articulation of the possibility of exceptional novelty subtracted from the order of knowledge. Badiou achieves this goal through the notion of the 'event', a singular happening absolutely indiscernible from the situation wherein it surfaces, and as such breaking the descriptive limits of ontology. The event, for Badiou, breaks the rules and the count of the situation, and produces a radical reconfiguration of knowledge: ontology lacks the discursive resources to reinscribe the evental irruption in its parameters. In order to explain the genesis of the event, Badiou exploits the difference between what he terms *presentation* and *representation*. Representation, Badiou explains, is an additional count, a 'count of the count' or survey which takes place after the initial count-as-one. In set-theoretic terms, representation is the process of formation of the power-set of a given set X, where each *element* belonging to X is combined along with all the others to become a *part* included in p(X). The *situation* quantified by X becomes, when represented, a *state* or *metastructure*. This second, metastructural reduplication is necessary, he explains, to secure the consistency of presentation from the errant, excessive inconsistency of the omnipresent but unpresentable void,[27] by recounting every element into parts of subsets. Structure and metastructure (situation and state) are, Badiou insists, radically different, and the gap between them is a 'permanent question for thought, an intellectual provocation of being' (2006a: 84). Badiou here catalogues different elements of situations according to their being both presented and represented (*normal* terms), presented but not represented (*singular* terms) or represented but not presented (*excrescent* terms).[28] Crucially, the theorem of the point of excess states that, for infinite sets, the power-set/metastructure *immeasurably exceeds* a given set/situation and therefore 'the "passage" to the set of subsets is an operation in *absolute* excess of the situation itself' since 'there is always ... at least one element ... of p(a) which is not an element of a. This is to say, no multiple is capable of forming-a-one out of everything it includes' (2006a: 84, 85). The excess between presentation and representation is a crack opened on the surface of ontology, from which the inconsistency of multiple-Being can unexpectedly ooze out.

Applying the insight of the theorem of the point of excess to the above typology of situations, Badiou approaches the event by posing its possibility within singular multiples which he terms abnormal sets[29] or evental sites: presented but *not* represented in the situation, *none* of the evental site's elements are presented in the situation. Such a site is *'on the edge of the void'* (2006a: 175), that is, it is extremely close to the complete unpresentability of the void itself. The event can be made possible by these singular multiples, composed by both elements of the site and, breaking the axioms of set theory (the laws of Being) *itself*. The central point that Badiou wants to bring home here is that the event's belonging to the situation is *'undecidable from the standpoint of the situation itself'* (2006a: 181): the decision of its belonging can only be taken as a wager through an *'interpretative intervention'* (ibid.). The mathematisable order of presentation lacks the resources to identify the possibility of conceptual novelty; set-theoretical ontology does not admit the event (which due to its self-belonging is in explicit contravention of the Axiom of Foundation)[30] whose insurgent *puissance* derives from unpresentable inconsistent multiplicity. Only a voluntary, subjective (and, as such, *aleatory*) effort can enforce interventional activity (formally authorised by the axiom of choice and taking the shape of a series of enquiries) within the situation in accordance with the event, by *naming*[31] it and placing oneself under its ongoing revolutionary injunction. The humanist/phenomenological prioritisation of subjective experience expelled by Badiou in his description of set-theoretical ontology (a purely formal ontology insensitive to content and meaning of the presented) is, ironically, re-injected into his account of how an event triggers a subject-interpellating truth procedure. As the impasse of ontology – the necessary incommensurability between the cardinality of situations and that of their metastructures – made it necessary 'to tolerate an almost complete arbitrariness of choice, that quantity, the very paradigm of objectivity, leads to pure subjectivity' (2006a: 280), the impossibility of lawfully discerning the event calls for the intervention of subjective thought.

Subject-driven change is the theme which Badiou has single-mindedly pursued since his early work, arguing that 'in science as in politics, it is the unperceived which puts revolution on the agenda' (in Hallward and Peden 2012: 198), that is to say, the unprovable conceptual alternative forces the scientific or political subject to embark in the risky pursuit of a new framework of thought: 'the new can only be thought as process' (Badiou in Riera 2005: 253), a process *embodied* by the militant.[32] Badiou explains that 'the names used by a subject – who supports the local configuration of a generic truth – *do not, in general have a referent in the situation*' and that

> [w]ith the resources of the situation, with its multiples, its language, the subject generates names whose referent is in the future anterior: this is what supports belief. Such names 'will have been' assigned a referent, or a signification, when the situation will have appeared in which the indiscernible…is finally presented as a truth of the first situation. (2006a: 398)

This, I want to argue, can be read as nothing but a re-description, cast in Badiou's vocabulary, of the kind of abductive inferential process I described in Chapter 3.[33] A truth, for Badiou, is indeed an infinite process of reckoning seizing the subject: the immanent construction of a generic multiplicity out the resources of the situation itself, triggered by the surfacing of an aporetic blind spot, or anomaly, within it.[34] A generic multiplicity is then

> a subset of the situation that is not determined by any of the predicates of encyclopedic knowledge; that is to say, a multiple such that to belong to it, to be one of its elements, cannot be the result of having an identity, of possessing any particular property. (2006b: 154)

Badiou, taking upon himself the task of renewing the Platonic[35] cleaving apart of knowledge from opinion against relativist perils (from Protagoras' individualism to Wittgenstein's conventionalism), reactivates the concept of truth by transforming it: a truth, by definition, cannot be *known* as true. No immediate justification, consonant with the finite parameters of the current (intersubjectively recognised) situation, can be offered and only *ex post facto* is it possible to offer an account of its structure.[36] The *obvious* is the rival of the *true*. As Bachelard would have it (if in more robustly realist register than Badiou's): '[r]eality is never "what we might believe it to be": it is always what we ought to have thought' (2002: 25).

However, *pace* Badiou, a fundamental asymmetry between the scientific and the political truth-procedure must be highlighted: Badiou's 'faithful subject' in physics or mathematics carries forward a truth-procedure as the asymptotic pursuit of some objectivity, and not as the gradual and *free* creation of a new, better *status quo* (as in politics). While there are no truth-makers in politics, the realist wants to hold that there *are* truth-makers in the universe, which cannot be subject to the kind of normative assessment which is paramount in political decisions. While both scientific and political (as well as artistic and amorous, to cover all Badiou's fields of truth-production) procedures are pursued by a committed subject (and it is crucial to track which *a priori* frameworks such subjectivity brings into scientific practice, as I noted in Chapter 3), the scientific procedure alone is *constrained* by its object matter, that is to say, only in science is there the possibility of *error*. To resort to hackneyed examples from the history of science, both phlogiston and the *élan vital* were, at some point, introduced into a body of knowledge as some unverifiable, non-veridical terms: however, the truth procedures associated with these terms eventually failed, because they not only had no referent within the contemporary scientific milieu, but had no referent in the physical world. Compare Badiou's own example drawn from the scientific procedure of Newtonian astronomy: he asks the reader to imagine astronomers postulating an as yet unobserved planet on the grounds of anomalies detected in the orbital trajectories of the known ones. Considering the scenario when a planet is in fact not observed, Badiou writes:

> [i]n [this] case, the relation [between the statement 'something is inflecting the trajectories' and the term 'planet', belonging to the vocabulary of the scientific situation] exists (expert calculations allow the conclusion that this 'something' must be a planet); but I do not *encounter* a term within the situation that validates this relation. It follows that my statement is 'not yet' veridical in respect to astronomy. (2006a: 402)

Generalising this scenario, he explains that, on the other hand

> [*i*]f I encounter this term, and it turns out to be connected to the name of the event, that is, to belong to the indiscernible multiple-being of

> a truth, *then,* in the situation to-come in which the truth exists, the statement will have been veridical. (2006a: 403)

This description can be interpreted as an abductive inferential procedure (indeed the alternative term *retroduction* seems especially apt in this context, highlighting the future-anteriority of discovery), where an unobserved entity is called upon as best explanation for the observational anomalies. However, Badiou's optimistic voluntarism fails to consider the (quintessentially scientific) predicament of failure or lack of confirmation during the self-correcting process of inquiry. This is the most noteworthy difference between Badiou's evental recasting of the situation in new, previously indiscernible terms and Bachelard's 'total reorganisation of the system of knowledge' (2002: 26) commanded by the historical overcoming of epistemological obstacles. Unlike Bachelard, Badiou seems to underplay the role of errors, miscalculations, misleading metaphors and incorrect solutions. These are normal and indeed necessary steps in the exigent progress of creative discovery: conflicts, inconsistencies, gaps and other cognitive obstacles that have a very real epistemological efficacy.[37] The scientific inquiry gains epistemic traction by proceeding along a multiplicity of paths occasionally discovering them to be dead ends – thus revising its advance accordingly. Badiou's own vision of the history of science (if anything worth this name can be found) is highly idealised, punctuated by events interpellating quasi-heroic subjects who single-mindedly pursue truths without faults or hesitation. So, the Badiouian template can be applied to the discoveries of Neptune and Pluto, both predicted via calculations grounded on Newtonian gravitation, and indeed subsequently observed (and to many more recent theoretical predictions in particle physics) but not, for example, to the notorious case of the planet Vulcan, the hypothesised planet alleged to be responsible for anomalies in the precession of Mercury's perihelion. Not only could Vulcan not be observed: Vulcan was simply *not there*. No matter how many inquiries the scientific subjects engaged in this truth procedure might undertake, the planet itself could not be forced into existence. Badiou's all too ideological predilection for *novelty* is unjustified: in heuristic processes, not all that is new is *necessarily* an element to be retained – new ideas can be plugged into theories and modify our conclusions, but all such new information can itself be subject of disconfirmation.

This, I believe, demonstrates how Badiou's 'truth', at least in the scientific generic procedure, cannot be taken to *replace* a more standard and realist understanding of truth as correspondence, but should be understood to describe a (fallible) knowledge-*ampliative* procedure (and not a *property*) propelling the process of theory-change. Badiou's 'truth' sits between a realist understanding of truth as something obtained thanks to a verification-transcendent, mind-independent state of affairs and an anti-realist stance where truth is an epistemic notion subordinated to warranted assertability;[38] his truth both *can* and *cannot* outrun possible justification. Wiggins correctly notes vis-à-vis the Peircean pragmatist notion of truth (but the insight applies to Badiou's truth equally well) that it is possible to vouch for an *elucidation* rather than a definition of truth, and that '[t]o elucidate truth in its relations with the notion of inquiry, for instance, as the pragmatist does, need not...represent any concession at all to the idea that truth is itself an "epistemic notion"' (in Schantz 2002: 318). On the one hand, Badiouian truth transcends verification to the extent that it cannot be procedurally pursued according to the established rules of a situation; on the other, there is no truth-maker 'out there' that makes a determinate statement true, but the truth of any claim is always the epistemic process of forcing that new truth into the situation.[39]

These shortcomings notwithstanding, Badiou's heterodox understanding of truth should not be dismissed as hopelessly wrongheaded, or useless, but should instead be located within the intellectual climate of late twentieth-century French thought, pervaded by a concern with the paradoxical articulation of absolute *novelty*,[40] struggling to navigate the dialectic between inseparable couples: inside and outside, structurally embedded meaning and structure-evading reference, closure and transcendence, semantics and syntax. Badiou seeks a truth that is not a static epistemic notion but an infinite ontological opening: '*the true only has a chance of being distinguishable from the veridical when it is infinite*. A truth (if it exists) must be an infinite part of the situation, because for every finite part one can always say that it has already been discerned and classified by knowledge' (2006a: 333). We can call this Badiou's *epistemic infinitism*: the laborious and *endless* creation of new representational resources to justify present hypotheses and allow previously unthinkable thoughts.[41] In Badiou's immanentist philosophy, secularised infinity features in both an ontological register (nature is infinite and non-totalisable) and in an epistemological one: the cognitive tracking of truth proceeds through an infinite iteration of reasons, without transcendent terminus. The rational creative inferential processes which allow finite thought to overcome its own conceptual limits are infinite to the extent that their objective is only asymptotically reached (due to both our cognitive closure and the ontological makeup of the universe). We are never done responding to the infinite *formal demands*[42] of reason-giving.

Badiou will formalise this infinite generic procedure of truth, the mediating process between the stasis of ontology and the dynamism of change, by exploiting Paul Cohen's notion of *forcing*, a mathematical procedure rigorously proving the possibility of constructing a proof for the truth or falsity of an undecidable proposition by immanently deploying an extension of a situation wherein such a proposition is either true or false. Specifically, forcing was devised by Cohen (1966) to prove the independence of the continuum hypothesis (and the axiom of choice) from the axioms of ZF set theory. Cohen proved that the *negation* of the continuum hypothesis is consistent with the axioms. Since Gödel had already established that the continuum hypothesis is consistent with ZF,[43] the two results implied that CH is independent and undecidable with respect to these axioms. However, as a piece of mathematical technology it can be generalised to deduce other conclusions. Cohen, Badiou explains, demonstrated that 'it is possible, in a quasi-complete fundamental situation, to determine under what conditions such and such a statement is veridical in the generic extension obtained by the addition of an indiscernible part of the situation' (2006a: 410), by creating a new model from a given countable model through the supplementation of a *generic* set. This is a set such as, for every predicate defined by the situation's rules (the 'encyclopedic determinant'), at least one element of the set will fail to satisfy *any* of them. As Hao Wang reports, Gödel qualitatively summarised the meaning of Cohen's procedure by saying that '[f]orcing is a method to make true statements about something of which we know nothing' (Wang 1996: 252). The ontological lesson Badiou draws from Cohen's proof is that there indeed is a radical incommensurability between infinite sets and their power sets, given that a model of ZF set theory can be built where the powerset of the denumerable cardinal $\aleph_0$ can be *any* other transfinite cardinal whatsoever, no matter how unthinkably large, $\aleph_1$ just as $\aleph_{273}$. The impossibility of proving the CH, then, amounts to the impossibility of prohibiting the event whilst preserving its indiscernibility to ontological standards, given the latter's reliance on the ontological impasse between presentation and representation, an indiscernibility that only a militant procedure can *decide*, via the construction of a generic set. The commitment to the truth-bearing efficacy of the forcing procedure is what keeps Badiou's own favoured *generic* orientation of thought apart from the *constructivist* and *transcendent* ones: truth can be ascertained outside of a paradigm of either rule-bound algorithmic progression or of divinely granted grace. The truth engendered by forcing is an immanent generic extension of a situation.

The main problem my interpretation has to face is this. Badiou's most pressing concern, and the ultimate motivation of his entire theoretical effort, is to prove that radical change is possible. His rationalism makes it possible for him to offer a general theory, through set-theoretic tools, of event-triggered change, which (1) encompasses both the conceptual and the real by taking Platonism as demanding the conflation of the two registers (the laws of thought are the laws of Being) and (2) is made operative by an act of subjective commitment, militantly pursuing the procedure of truth. I (like any realist) on the other hand, defend the intransitivity of the universe, reject the equation of thought and Being and any ontological generativity of subjective will, and therefore want to recast the notion of event as triggering a rational (and defeasible) procedure of abductive discovery, as opposed to a subjective procedure of constitution, the success of which is guaranteed by sheer fidelity. The defeasability is entailed by the non-monotonic nature of abductive inferences: in Badiouian terms, we must be open to the possibility that the result of a future inquiry will invalidate our truth procedure. In sum, in order to reconcile structural realism and Badiouian ontology, attempting to resolve the incompatibility between something real as an algorithmically projectable structural pattern and the non-algorithmic nature of Badiou's generic truth procedures (a problem expressed in the passage by Fraser quoted above), it is necessary then to deflate Badiou's notion of an *ontology*-shattering event, reinterpreting it as the catalyst for an *epistemic* procedure puncturing the situation of *knowledge*.

So while I claim, with Badiou, that there is more to the universe than can be projected via a strictly conceived algorithmic model, I nonetheless argue that the truth-process need not be triggered by an evental leakage of Being *qua* inconsistent multiplicity at the juncture between structure and metastructure (since I have already rejected the necessity of referring to inconsistent multiplicity at all).[44] I once again shy away from the vestigial Heideggerianism detectable in some of Badiou's discussion of truth:

> [i]nasmuch as the unfathomable depths [*fond sans fond*] of what is present is inconsistency, a truth will be that which, from inside the presented, as *part* of this presented, makes the inconsistency – which buttresses in the last instance the consistency of presentation – come into the light of day. (1999: 106)

The truth-procedure as delineated in my reconstruction of Badiou's epistemic infinitism does not require this ontological (un)ground for its success, but is an immanent process for the overcoming of knowledge/relative ignorance. Unchained from its origin in an ontology-breaking event, the truth procedure represents an inferential path for a rational agency decoupled from a naïve notion of randomness as (Sartrean) voluntaristic freedom, and can be considered *productive* in a non-ontological, heuristic sense. A hypothesis is generated, which will be forced into truth through the rigours of inquiry. At the same time, the strict constraints imposed by OSRists on the range of acceptable heuristic procedures in science-driven metaphysics need to be relaxed (so as to include methods of creative reasoning exceeding the resources provided by classical logic, that is not reducible to the application of deductive or inductive algorithms) and the strict alliance between being and knowledge presupposed by it rejected. While a certain amount of speculative reasoning should be admitted in scientific thought (both the natural sciences and mathematics), it is a rational imperative to envisage the *process* of conceptual change and invention[45] as a procedure that can be made explicit and which is not subordinated to a singular, sudden and quasi-miraculous *act* of personal genius: the context of discovery is not impervious to logical reconstruction (and the study of the cognitive limits of human beings can be seen to map the limits of creative discovery), but is a moment of the infinite progress of rationality towards the regulative ideal of a limit science.[46] It seems to me that Badiou's militant, inquiring subjects, are then best seen, in a Peircean light, as those 'followers of science ... animated by a cheerful hope that the processes of investigation, if only pushed far enough, will give one certain solution to each question to which they apply it' (*CP* 5.407).

In sum, the *extra-ontological* role the event plays in Badiou's system can (and must) be eliminated together with any reference to inconsistent multiplicities. The possibility of radical novelty and genuine breaks in knowledge introduced by the Badiouian 'event' can be preserved once the 'militant' or 'faithful' subjectivity that, in Badiou's account, voluntaristically brings about novelty and forces truth into a given situation is replaced with the idea of rational inferential processes of discovery *about* but not *detached from* the natural order. It is greatly to Badiou's credit to have restored to (continental) philosophy the task of an uncompromising pursuit of truth, against all sophistic foreclosures of it in the name of linguistic relativism or existential finitude.

After having criticised so much of Badiou's thought it might be useful to highlight how his resolute rehabilitation of opinion-transcending truth against the constructivist sovereignty of language is the *second* trait for which I take Badiou as a highly significant figure in contemporary philosophy. The first, I reiterate from Chapter 1, is his articulation of the implicit secularising or downright anti-theological value of post-Cantorian mathematics. Naturally, these are two sides of the same rationalist coin: to uphold the infinite power of reason against the constraints of finitude. As such, his conception of infinite truth procedures, forcing creative thought out of the rut of worn-out repertoires of knowledge, is a precious resource for those realists who, as I explained in Chapter 3, consider it part of the realist's task to promote the abductive exploration of new concepts. However, the notion of a truth-generating event, indiscernible from the point of view of the ontological situation *qua* mathematical discursive articulation of the status quo becomes unintelligible once we identify (mathematical) structure with reality itself. Nothing can break out of mathematical articulation to the extent that, the naturalist insists, nothing either transcends or subtractively underlies reality. Where Badiou claims that 'being qua being does not in any manner let itself be approached, but solely allows itself to be sutured in its void to the brutality of a deductive consistency without aura' and that it 'does not diffuse itself in rhythm and image, it does not reign over metaphor, it is the null sovereign of inference' (2006a: 10), one should wonder whether the rational power of inference (especially in its abductive form) really does need the supervision of a null sovereign inaccessible by any cognitive/epistemic route. Cognitive naturalism is achieved once reason (*qua* adequation to inferential rules) is conceived as an evolutionary achievement of sapient creatures *out of* (but not *reducible to*) non-thinking causal reality.

Regrettably, the price Badiou pays for his rejection of onto-theological transcendence is that of requiring a subjective intervention – capable of forcing situation-transcending truth out of the immanent constraints of ontological discourse – whose origin and motivations remain as inscrutable as those of a Deity (the formula 'militant fidelity' flounders in a misty political analogy).[47] Likewise, Badiou's 'event' appears totally unmoored from its socio-historical and rationally reconstructible triggers, unaccountably springing into a situation ready to be its operator of change, all too reminiscent, once again, of Heideggerian *Ereignis* and the 'appropriation' of subjectivity into truth which the latter signals. Bensaid indeed writes that '[d]etached from its historical conditions, pure diamond of truth, the event…is akin to a miracle' (in Hallward 2004: 101), and his indictment is echoed by Smith, who argues that '[t]hough Badiou is determined to expel God and the One from his philosophy, he winds up reassigning to the event, as if through the back door, the very characteristics of transcendence that were formerly assigned to the divine' (in Hallward 2004: 93).[48] We cannot envision the process of conceptual creativity as standing on the sands of this evanescent constellation of notions. New scientific ideas do not spring fully armed out of the scientist's forehead, but develop through constant and self-aware negotiation with our social and cognitive past, and involve the risk-taking pursue of a futural knowledge-state. As Lakatos had his 'Teacher' trenchantly aver in *Proofs and Refutations*: 'I abhor your pretentious "insight". I respect conscious guessing, because it comes from the best human qualities: courage and modesty' (1976: 30). The genesis of new concepts is to be explained through cognitive structures and rational capabilities which humans collectively instantiate, not via some inexplicable flash of other-situational insight in the mind of a quasi-heroic individual. All that is needed is the courage to pursue an infinite truth and the modesty to recognise that one's reason-giving might be always in need of revision.

In order to defend a demystified account of Badiou's conception of truth, we should follow a path extending from Sellars's work (and its Kantian roots)[49] through Brandom's inferentialist account of rationality and the spirited (yet qualified) advocacy of both by Brassier, and endorse a 'non-naturalisable' (as *logically* independent from its ontological basis) rational subjectivity which has, however, firm neurobiological roots on the continuum of natural causes.[50] The rational subject gripped by (socially instituted) inferential norms of concept-use is the agent of instantiation of the process of truth-discovery: the break between accredited knowledge and exceptional truth sought by Badiou (and any realist, in my reckoning) need not be equated with an aleatory jump from inconsistent Being to an otherwise-than-being enforced by a selfhood impossible to assimilate within a naturalist worldview. On the contrary, it should be reconceived in terms of the trackable operations of a naturalisable yet norm-bound reason capable of conceptual creativity.

5.4 Metamathematics and randomness

Having deflated Badiou's truth of its ontological condition of possibility in inconsistent multiplicity, and argued for its epistemic understanding as a limitless abductive process of creative inquiry, I must now return to the ontological register, and consider how Badiou's commitments to Being as pure multiplicity and his assumption of metamathematical results as 'laws of being' can survive the matherialist collapse of 'being' with structure. How to reconcile the two domains I have now prised apart, that of rational heuristic procedure, on the one hand, and the objective structure of reality, on the other? This question, of course, bears on another one: How can the structuralist ontology of patterns and Badiou's ideas about a non-totalisable reality and about the production of truth be reconciled?

As I mentioned in Chapter 4, Ladyman and Ross pose as a condition of existence for a 'real pattern' that it be algorithmically *compressible* and *projectable* to future states. Badiou, on the other hand, warns us against the constructivist orientation of thought underpinning these information-theoretic notions which 'sets forth the norm of existence by means of explicit constructions' and 'ends up subordinating existential judgment to finite and controllable linguistic protocols. Let us say any kind of existence is underpinned by an algorithm allowing a case that it is the matter of to [sic] be effectively reached' (2006c: 55). 'In short', Badiou concludes 'the orientation of constructivist thought subsumes the relation to being within the dimension of knowledge... that which is not susceptible of being classified within a knowledge is not' (2006a: 293). Badiou would see the pattern-ontology provided by structural realism as foreclosing the possibility of novelty through the enforcing of generic additions, being subjected to the closed rules of algorithmic processes and thus deprived of the reservoir of contingency/inconsistency affording evental irruptions. Only the combined effect of an *uncomputable* event and a *subjective* militant procedure can, for Badiou, produce any truth. However, I have already stressed how, in my account, both these elements fail to pass a naturalist test and tend to reintroduce ineffability into our ontology, and indeed Badiou's hostility to computational procedures in favour of the alleged uniqueness of human capabilities might be downright incorrect and narrow-minded.[51]

Ray Brassier has highlighted the unsophisticated understanding of subject-generated randomness Badiou seems to espouse and suggested a possible way out of his voluntarism by referring to Gregory Chaitin's metamathematical work on algorithmic information theory, offering a way to 'undermine the distinction between the subjective measure of excess and its objective calculation, and to determine the abyss between the finitude of truth's subjective act and the infinity of its generic being' (in Hallward 2004: 56). This is a promising suggestion as indeed there is nothing in Badiou's thinking (his 'human exceptionalist' hostility to computational methods notwithstanding) that cannot be receptualised in terms afforded by algorithmic computations: computational models of creative reasoning are far from inconceivable,[52] and since Turing's work[53] we have learnt to demystify (but, notice, not trivialise) the power of human thought, since no form of reasoning instantiated by us cannot in principle be replicated by a computing machine, rational agents being no more than situated physical systems and thus bound by precisely defined computational (and *physical*) limits,[54] (this precept grounds the research into Artificial General Intelligence (AGI): a general, human-like form of intelligence – capable of creative and explanatory reasoning – in principle replicable by artificial systems). In other words, the aleatory character of Badiou's procedures can be captured and made intelligible by computational models. And, the matherialist adds, it is *reality in itself* that *imposes* physical limits to what can be computed while *affording* the immanent randomness required for creative processes. To expect an intrinsic indeterminateness of reality, against the cosmic harmony of the ontotheological worldview, is a *realist* postulate and not an anti-realist exhortation to scepticism: randomness and indeterminacy can be rationally appraised and do not invite the bleary demise of reason. The insight that nature is a domain where regularities and order are the exception follows (as I have noted already in Chapter 3) the recent questioning of the image of ordered science and its reliance on immutable laws of nature as explanatory and predictive mechanisms, initially spearheaded by philosophers of science of the Stanford School and now a widespread and vigorous research programme (and adumbrated by Peirce's speculative cosmology, under his notion of *tychism*). I find William Wimsatt's vivid description shrewdly captures the general perspective guiding this project:

> [i]nsisting on an exact, precise, complete, exceptionless description can hide important order that is there. Sometimes, if we are willing to tolerate a few exceptions, context-dependencies, and approximate truths – if we are willing to 'defocus' a little – we can get nice, compact, *ceteris paribus* qualified but robust generalizations. That's still worth doing in the vast regions of our conceptual world – most of it – where that's the best we *can* do... It's also worth doing because it opens our world to a deeper explanation, for constrained fuzzy order is the general case: *on our time and size scales and for several orders of magnitude in each direction, it is exceptionless crisp precision that is the degenerate special case. Nor are we – detecting and assessing this order – disembodied cognitive engines: we yoke cognition and affect quasi-independently in a noise-tolerant manner in the service of local and more global ends.* (2007: 33)

What must be underlined is that the matherialist thesis allows us to *both* accept a non-reductionist theoretical unification of science under the ultimate authority of mathematical description (via a scale-relative ontology of patterns) *and* have rational grounds to defend a certain 'disunity' of nature (by endorsing metamathematical, anti-foundationalist results as bona fide descriptions of reality). Without forfeiting any kind of order (and with it, the hope of regulated scientific inquiry), metamathematical results demonstrate how structural organisation immanently produces self-imposed limits and paradoxes.

Metamathematics, as the employment of mathematical methods to examine the large-scale structure of mathematics and probe its limits, originated with Hilbert's foundational project: to confirm and secure the power of mathematics by defining the parameters of a formal axiomatic system capable of deductively producing the entirety of mathematical theorems, and proving them via an unequivocal decision procedure (in Badiouian parlance, Hilbert's ambition was to impose a metastructure on the mathematical situation). Throughout the twentieth century, however, metamathematics has mostly become a reflection on the intrinsic limits of mathematics and the paradoxes immanent in mathematical structures. Metamathematics is where mathematics takes on the project of demarcating its own boundaries. Chaitin[55] places himself in the tradition of metamathematical research which shattered the Hilbertian imperative of total knowledge, from Gödel's incompleteness theorems, forcing a choice between the system's completeness or consistency, through Turing's work on computable functions, demonstrating the impossibility of resolving the *Entscheidungsproblem*.[56] Through his notion of program-size complexity Chaitin redefined the concept of randomness as *incompressibility*: a random number is one that no algorithm or program of a shorter number of bits than the number itself can produce, just as a random set of data is one that no theory can reduce to a more economical generalisation.[57] More formally put, the information in a given string of digits is equal to the length of the shortest formulation of a program that would lead a universal Turing Machine to output that string.[58] Chaitin, extending Gödel's explorations of the limits of formal systems, proved that (within any formal axiomatic system) there can be *no* formal proof that a given program/theory is 'elegant', that is, is the smallest possible program (in terms of bits or binary expansion) that can produce/explain a set of data. Conversely, there will always be numbers whose randomness cannot be proved, in fact the majority of them. He proceeds to show how there can be no algorithm to solve Turing's halting problem, since, if there were such an algorithm, that could be used to find all the elegant programs (being able to pick out the programs that halt, one would then be able to check their outputs and get rid of all but the elegant ones).

Instead of asking, as Turing did, whether any given program will halt, Chaitin looks at the ensemble of all possible computer programs and asks what is the probability that a given program chosen at random, running on a universal Turing machine, will ever halt (0 if no program halts, 1 if all programs halt). This halting probability is expressed by a number he labels Ω, a real number, in the form of a binary expansion, between 0 and 1. Ω, Chaitin demonstrates, is utterly uncomputable, as each of its bits is as algorithmically random as an independent toss of a fair coin, and no pattern whatsoever can recapture its complexity (that is, the smallest program to output the first N bits of Ω is also N bits long). Ω can only be followed, starting with its lower limit, bit after bit without having any way to explain why the bit just encountered is thus and so, and without ever reaching its upper limit. Ω is then a real specifiable number which defies the Leibnizian PSR[59] since there is no reason (shorter explanation) why each of its bits is as it is (but, notice, not the 'epistemic', revised Spinozist-naturalist understanding of the PSR I have offered above: Ω can be asymptotically pursued, step by step, but cannot be given a top-down divine reason for its being what it is).

Chaitin argues that the randomness (algorithmic incompressibility) displayed by Ω in a relatively basic mathematical field like number theory demonstrates that it is an ineliminable feature of mathematics at large. He develops this insight into the ubiquitous nature of randomness by marrying it with a computationalist, digital metaphysics, the view that the universe is made by – or can be reduced to – information,[60] thus arguing that the overwhelming majority of this information/reality is not accessible because incompressible, its brute facticity impossible to recapture in more general and economical laws or *explanations*. An explanation needs to be a shorter or simpler string of bits (or words) than the phenomena or data it purports to explain. If understanding, then, is compression, as Chaitin argues, incompressible information can be accessed, but not understood. Randomness then 'is as fundamental and as pervasive in pure mathematics as it is in theoretical physics' (Calude and Chaitin 1999: 320) and the two disciplines are on an equal footing when it comes to investigating their domain, where most structures are true for no reason, that is no reason *simpler than they are.* Facticity *qua* incompressibility affects *every* elementary existent. In broader epistemological focus we can say, in Leibnizian terms, that the vast majority of truths in the universe are contingent as requiring an *infinite* process of *finite* analysis: without a God capable of intuiting the end of analysis (thus guaranteeing the truth's truthfulness) it is the task of cognisant humans to pursue *discursively* infinite truths (and it is this formal infinite exceeding of our epistemic grasp which underpins realism). This formalised notion of randomness, then, can be used as a corrective to Badiou's vaguely formulated reliance on subjective contingencies; as Brassier rhetorically asks: '[d]oesn't the evental excess which, for Badiou, indexes the inconsistency of the Real and necessitates the uncomputable freedom of axiomatic decision, find embodiment in Ω's objective randomness at least as legitimately as in subjective intervention?' (in Hallward 2004: 57). In the light of a structuralist ontology of patterns, then, Chaitin's insights into randomness indicate that what is algorithmically compressible and projectible (that which is 'real' according to this ontology) does *not* exhaust what exists. Badiou's claim that the knowable does not exhaust the real is at once vindicated and undermined, since it is through computational inquiry into the objectivity of mathematics (and not a subjective wager) that we reach this anti-constructivist conclusion.

It is here that I am in broad agreement with Paul Livingston's revision of the consequences of the Cantorian secularisation of the infinite heralded by Badiou. Livingston (2011, 2013) draws from Gödel the insight that two possible consequences follow from Cantor's and Gödel's own metamathematical work: one insisting on an inherent, non-formalisable capacity of human minds to exceed algorithmic processes (Gödel's and Badiou's position) and the other more humbly acknowledging the existence of *well-defined* problems which, however, are constitutionally unsolvable. Livingston opts for the latter stance, which he labels the 'paradoxico-critical' stance, amounting to 'an inexhaustible inscription of the undecidable as such ... in the very structure of mathematical reality' (2013: 106), thus preserving inconsistent completeness rather than consistent incompleteness. The merit of this exclusion of the unaccountable power of subjectivity to grasp (or produce) truths non-discursively/mechanically, and of the preservation of a complete (that is, gap-free) but inconsistent (dis)order of Being, is that of respecting immanentist strictures. And as Livingston notes, echoing my remarks against Badiou's hasty dismissal of the Kantian project in favour of a rationalist omniscience,

> besides being more obviously compatible with materialism because not in any way at odds with mechanism, the paradoxico-critical outlook makes it possible to preserve an outlook and practice that continues the classical orientation of criticism with respect to the capacities and practices of the human subject, in the altered conditions post-Cantorian thought. (2013: 107)

Livingston defines his broader position as a 'metaformal realism', a realism interested in the relation between our human capacities and the infinite thinkable structure which either horn of Gödel's distinction implies, since

> each term of Gödel's disjunction reflects the necessity, given Gödel's theorems, that any specification of our relevant capacities involve their relation to a structural infinity about which we must be realist, i.e. which it is not possible to see as a mere production or creation of these capacities. (2013:101)

These reflections upon the *formalisation* of immanent limits and aporiae in formal systems present, as a consequence, the existence of the undecidable or unknowable as that which cannot be reinscribed in a concept but needs to be asymptotically pursued by endless self-correction. This is warranted, from the matherialist perspective, not only by our epistemic constraints but also by the very structure of matherial reality. The former are constantly yet endlessly transcended – and this is the domain where conceptual creativity helps us update and improve our conceptual frameworks and where abductive processes regulate the selection of new conceptual possibilities for maximal explanatory output. The latter can be seen as the intrinsic, uncircumventable *yet not mysterious* feature of reality which allows us to counter both the (Dummettian or mid-Putnamian) anti-realist notion that all truths must be knowable (there are mathematical facts independent of our capacities to grasp them) and a structuralist totalisation (is no knowable meta-principle guaranteeing the consistency of structure).

Reason can conceive of unanswerable questions which remain infinitely answerable through the process of reasoning. So rational-computational abductive processes of truth-pursuit cannot meaningfully be said (with Badiou) to 'disrupt' or 'break' ontology since, the naturalist matherialist holds that after the metamathematics of Gödel, Turing and Chaitin no qualitative gap can be drawn between the capabilities of human-instantiated rational thought and algorithmic computation, and that the phenomenon of randomness is part of the immanent mathematical universe (where inconsistencies and contradictions are the norm), not the result of an unfathomable non-ontological break.

Conclusion

In order to conclude this book, I shall now reconstruct my main line of argument.

Opening with an appreciative reconstruction of Badiou's mathematico-ontological project, I identified what, by my naturalist reckoning, is its most glaring issue: the lack of a proper account of the way in which ontological, set-theoretical 'situations' come to have a relation with/effect upon non-ontological, empirical ones. Bereft of clear empirical purchase, Badiou's ontology can function, at best, as an analogical model for his ideal kind of militant political practice, but remains unable to offer any insight into non-human realities. I argued that Badiou's rationalist understanding of mathematics as founding the pure laws of thought is a questionable outcome of his equation of Being and thought, and that indeed mathematical abilities can be given a demystifying cognitive-evolutionary account, where the thought of mathematics is shown as emerging from the pre-noetic Being of nature. Moreover, the quandary of the applicability of mathematics to the physical world can be given other, more empirically grounded explanations than Badiou's own non-explanation. On the way to a full explication of a possible answer to this problem, I offered some reflections on the explanatory task of contemporary realism and my own interpretation of a responsible yet 'transgressive' naturalism, aiming at (speculatively) supplementing naturalist orthodoxy without being lured into the 'liberal naturalist' reintroduction of non-naturalisable domains. My proposed matherialist resolution of the mathematics–physics relationship emerged, then, from the creative integration of the metaphysics of Ontic Structural Realism with a structuralist view of mathematical entities, rejecting the dualism of concrete and abstract and presenting a thoroughly immanentist and scale-relative ontology of structures 'all the way down'. I argued that such an ontology does away with the 'depth' necessary for Badiou's (and Heidegger's) project of fundamental ontology and – having accordingly recast Badiou's account of truth as a rationally (and computationally) accountable process of abductive discovery – I noted how it can be, in turn, supplemented with meta-mathematical insights into the intrinsic features of structures. These allow us to reintroduce a Badiouian insistence on the secularising power of mathematics and on the novel character of truth by demonstrating, against onto-theological views, how metamathematical inquiry offers formal legitimacy to the inconsistency and contingency proper in structural systems and therefore (the matherialist is entitled to infer) of reality itself.

* * *

One of my core and recurrent claims throughout this book is that the naturalist is committed to ontological and methodological immanence: the former commits us to defending the continuous nature of reality, wherein all relations pertaining to that which is real are always internal relations, and never relations to something external (be it theistic creation, Platonic participation or any other form of vertical metaphysical dependence). There is nothing outside, beyond, against or before what there is. The latter commits us to the assumption that successful epistemic practices in one field of knowledge (natural science) should be just as successful in any other field (for example, mathematics), and vice versa. These are the commitments buttressing the matherialist position, which can thus be defined as nominalist with only partial accuracy. Recall Burgess and Rosen's memorable anecdote:

> [f]inally, after years of waiting, it is your turn to put a question to the Oracle of Philosophy. So you humbly approach and ask the question that has been consuming you for as long as you can remember: 'Tell me, O Oracle, what there is. What sorts of things exist?' To this the Oracle responds: 'What? You want the whole list? Look, I haven't got all day. But I will tell you this: everything there is is concrete; nothing there is is abstract. Now go away and don't bother me. (1997: 3)

My own Oracle would have answered, perhaps even more cryptically, that the first step towards the knowledge the petitioner seeks is to discard the abstract/concrete binary: this might be an expedient methodological distinction, but it is not a *real* distinction. In particular, the nominalist would want to reject the existence of sets, while my proposal here is precisely to preserve (pure) sets as building blocks that are neither abstract nor concrete (or both). This position, then, could be placed somewhere between the two different naturalisms of Quine and Sellars, and their divergent opinions regarding mathematical universals: the former's reluctant Platonism admitting sets into the list of indispensable ontological commitments and the latter's nominalist functional role account of abstract entities, proposed to counter the unwelcome (for a naturalist) epistemological difficulties raised by the reference to sets. While, with Quine, I accept sets as existents (if for very different reasons, and openly embracing a certain Pythagoreanism) I endorse Sellars' admonitions against re-introducing any 'radical discontinuity into [the naturalist's] Continuum' (1980: 16).[1]

I should stress that, as in the case of OSR, objections can be raised against this position. Both MS and OSR (and even more so their conjunction) are to be seen as research projects, and not as established, irrefutable philosophical truths. However, it seems to me that together they can offer a solid and coherent naturalistic worldview, one ripe for encounter with the meta-ontological reflections of Alain Badiou.

To my knowledge, (a variant of) this position has been explicitly defended only twice in recent times. First, by Randall Dipert, who proposed a holistic (purely relational) structural metaphysics where the world as a whole can be mathematically described by graph theory, doing away with intrinsically identifiable, self-subsistent relata. In this picture, physical objects 'themselves are composite entities, subgraphs of the world graph. Physical microstructure is graph-theoretic macrostructure' (1997: 356) and even '[t]houghts are also subgraphs of certain sorts' (ibid.).[2] Second, by theoretical cosmologist Max Tegmark (2007),[3] who proposed his mathematical universe hypothesis, claiming that 'our external physical reality is a mathematical structure' and that '[t]he various approximations that constitute our current physics theories are successful because simple mathematical structures can provide good approximations of certain aspects of more complex mathematical structures (2007: 104, 107). In such a mathematical structure human beings would belong to the group of 'self-aware substructures', generally endowed with an inside perspective (a 'frog view') expressed in natural language but capable of achieving an outside perspective (a 'bird view') if reasoning in purely mathematical terms. Importantly, Tegmark is ready to concede, as I am obliged to, that the equation of mathematical and physical reality entails that any mathematically possible structure should enjoy real existence in some (part? region? brane?) of the mathematical multiverse, hence dodging the problem of the overdetermination of physics by mathematics.[4] I am not sure that this claim is as sensational as it sounds: it seems to me that once the concrete/abstract distinction is surrendered, it commits us to nothing more than what we are already committed to when considering the cosmological idea that there might as well be an infinite plurality of multiverses – a speculative and contested thesis, but not alien to contemporary cosmology. It does, on the other hand, support those connections between the structural limits of computation and the boundaries of the real I defended in Chapter 5.[5]

The cautious and doubtful way in which the majority of philosophers of science voice their thoughts, when considering the possibility of a total erasure of the dualism between concrete and abstract (or physical and mathematical), is unsurprising (see for example French's very detailed yet singularly noncommittal engagement with Tegmark's thesis (2014, Chapter 8)). This thesis can be given at best a contextual or abductive justification (steering one's conceptual boat according to the theoretical virtues of ontological economy and immanence, defending the matherialist stance as the best explanation for the successful applicability of mathematics), an empirical proof being in principle foreclosed. These well-warranted empiricist or verificationist scruples should not, however, deter the philosopher seeking a synoptic view from taking one more speculative step than others are willing to take. All too often, it seems, realist positions are formulated as a response to the most cogent anti-realist arguments. While such a dialectical evolution of concepts is a necessary dynamic of intellectual progress, to subordinate realism to a 'conceptually prior' anti-realism turns the former into a rather irresolute 'anti-anti-realism'.

It is unquestionably true that many anti-realist objections cannot just be wished away, and that epistemological criteria needed to demarcate warranted claims from metaphysical flights of fancy should be held firmly in view. However, the realist should not commence her theoretical work on the defensive but should dare to speculate: to undertake a detailed reconnaissance of the scientific frontier and venture a Peircean abductive leap from the trampoline of our best empirical knowledge towards the as yet unknown features of reality and the very bounds of sense. It is here that a looser approach (one, if you wish, of more 'continental' pedigree) allowing and promoting the creation of new pathways in conceptual space can offer sizable rewards to the philosopher of science. The concepts thus created can then be available for 'plundering by philosophers of science in order that [they] might exploit them for [their] own purposes' as French and McKenzie (2012: 44) recently put it, defending a certain role for scientifically 'disinterested' metaphysics, thus helping the interpretation of the ever more puzzling and counter-intuitive features of reality.

I heartily endorse Ladyman and Ross's belief that the intellectual duty of 'extending the Enlightenment project' (in Ladyman et al. 2013: 113) is better discharged by walking the epistemically royal road of science and by endorsing '[t]he realist's epistemological optimism about the boundless future of scientific progress' (ibid.: 147). I find the revisionary metaphysics of OSR refreshingly ambitious and – for those interested in such cross-pollination – able to deliver a naturalistically justified (missing) link between Badiou's ontology and the empirical world. I do not, however, fully endorse OSR's post-Viennese verificationist strictures and its naturalism-driven rejection of the *a priori* and of a legitimate role for meta-scientific reflection. I would argue that perhaps an overly strict observance of physics-first criteria of judgement undercuts the speculative ambition that Ladyman and Ross see promoted by the mathematisation of our world-picture, and that critical attention to the historical evolution of *a priori* structures of thought (the kind of neo-Kantian project that Wilfrid Sellars and, more recently, Michael Friedman have differently pursued) will never be rendered totally redundant by naturalised metaphysics.

Indeed, I envision my future work as focusing on the construction of a synoptic philosophy of science informed by a conceptualist approach to epistemology, best explicated – once again in a cross-traditional key – through the work of Bachelard and Sellars. Such approach would be capable of reconciling attention to the historical development of conceptual frameworks (Friedman's 'dynamic Kantianism') with the naturalist ambition to construct a purely 'scientific' metaphysics and a robust scientific realism.[6]

I think that the metaphor of continuity, often invoked in definitions of naturalism (philosophy as 'continuous with' natural sciences) should be taken more seriously than it often is: it is *precisely* the naturalist vantage point that should make it clear to us that the natural sciences, philosophy and history are indeed on a continuum. Methodologically, different heuristic strategies might be adopted to best achieve explanatory purchase on the desired scale of phenomena, but these remain distinct yet continuous forms of *the same* intellectual enterprise, the human quest for knowledge in a universe of immanence.[7] The ideal of a synoptic vision should be kept in sight at all times: plural orders of explanations, including a relative autonomy of the ineliminable conceptual framework of the manifest image, can be defended even whilst final ontological authority is delivered to mathematical physics. A minimal inferentialist definition of knowledge as the ability to locate one's beliefs about the world in a nexus of conceptual justification together with a naturalistically mandated rejection of ontological fundamentality entailed by the patternist approach of OSR (whilst certain 'core' patterns are detectable at all scales, even those do not resolve into 'substantial' fundamental components ontologically prior to their constitutive relations) can help pursue the synoptic goal. These stances would allow us to preserve manifest image talk as knowledge-producing and (conventionally) true whilst recognising the (ultimate) truth of a description couched in the conceptual framework of the physical sciences.[8]

* * *

As I noted in my Introduction, the best outcome I could hope for this book, inasmuch as it seeks a synopsis of heterogeneous approaches, is internal consistency. I believe that I have achieved such consilience (if at times straining arguments beyond their intended domain of applicability – but such is the synoptic philosopher's trade) between the stages of my discussion. The practice of crossing disciplinary barriers, even limiting oneself to those within the field of philosophy, is a dangerous practice, as a lack of specialisation might lead to incomplete accounts, trivial insights or even downright incorrect conclusions. I have done my best to avoid these unwelcome outcomes. This, after all, is but a step in my own life-long self-correcting theoretical enterprise.

Notes

Introduction

1. Much ink has been spilled trying to assess the validity of these terms (were Carnap and Wittgenstein not born on 'the continent'?), to define the essential traits of the two traditions (argumentative style and logical clarity vs. oblique prose and historical consciousness?) and to trace the origin of the parting of ways (Carnap vs. Heidegger? Brentano's legacy? Frege's revolution in logic?). These are worthy enterprises, but it seems to me that more often than not the earnest focus on 'reconciliation' gets in the way of the more pragmatic strategy of acknowledging the *socio-institutional* division whilst rejecting its *philosophical* meaning. Hence my use of the word *neglect*. My sincere hope is to see, in a few decades, a new generation of philosophers taking over, having been educated by teachers who thoroughly neglect this division. For these young scholars the labels 'continental' and 'analytic' will be mere objects of historical curiosity, and they will look down upon those hard-headed philosophers still proudly self-identifying as 'analytic' or 'continental' as quaint, and somewhat amusing, relics of a bygone era.
2. I borrow this useful term, indicating proficiency in both continental and analytic methods and vocabularies, from Rorty (in Prado 2003).
3. By my reckoning, Moore is an exemplar synoptic philosopher, defining himself as a 'philosophical generalist' (2012: xviii–xix), and lamenting that 'the narrowness of focus that we see nowadays within philosophy poses a threat to its being pursued at all, in any meaningfully integrated way' (2012: xix).
4. Throughout this book I shall prefer the term 'universe' to 'world' or 'external reality'. It seems to me that, first, the term 'world' is an anachronistic piece of metonymic philosophical lexicon (see Cazeaux 2007, Chapter 6 for a discussion of the 'world' metaphor in epistemology). Yet it is still ubiquitously used in contemporary philosophy, often with ontologically extravagant connotations (see, for example, Goodman 1978; Lewis 1986; Gabriel 2011; Kuhn 2012). In our collective consciousness we have long since moved our abode from a finite world to an infinite universe, and it seems mistaken to employ the term 'universe' to refer exclusively to the infinite and star-spangled expanse of space beyond the confines of our planetary atmosphere, as this seems to reinstate an arbitrary form of dualism between our familiar 'homey' world and the alien, hostile universe. Moreover, the term, in its vagueness, lends itself to the kind of anti-realist, semantic deflation operated by the likes of van Fraassen, who invites us to interpret 'world' as 'a context-dependent term that indicates the domain of discourse of the sentence in which it occurs, on the occasion of utterance' (2002: 24). Second, the qualifier 'external' seems to me, again, to rely on a tacit dualism between internal and external which the naturalist simply does not acknowledge. There is but one reality, the very same reality both inside and outside of our skulls.

Additionally, I will refrain as far as possible from using the noun 'nature' in these contexts; doubtlessly one of the most semantically loaded terms in Western languages I prefer to use the no less complex but somewhat more circumscribable term 'reality'.

5. As Beiser puts it 'to historicize our thinking means to recognize that everything in the human world – culture, values, institutions, practices, rationality – is made by history, so that nothing has an eternal form, permanent essence or constant identity which transcends historical change' (2011: 2).
6. See, in particular, Hallward (2003); Norris (2009); Pluth (2010).
7. I take issue with the regulative role that politics all too often assumes in Badiou's ontological system-building. No good *realist* ontology is built upon ethico-political commitments (but a *successful* one can be: ask Paul of Tarsus). On the other hand, a truly emancipatory politics requires a suitably robust ontology, to offer a metaphysical map of reality for generations of militants to explore. I will very elliptically buttress this claim with a single word: Spinoza.
8. The term 'bold naturalism' is employed by McDowell (1996) as a foil against his own 'naturalism of second nature'. Liberal naturalism is a McDowell-inspired stance defended by a number of philosophers, articulated most clearly in two anthologies edited by De Caro and Macarthur (2004, 2010) and defined as a naturalism that 'offers a broader, more expansive conception of nature that makes room for a class of nonscientific, but nonetheless nonsupernatural, entities' thus occupying 'the typically overlooked conceptual space between Scientific Naturalism and Supernaturalism' (2010: 4, 9).
9. Thorough introductions to the arguments of *Logics of Worlds* are Hallward (2008) and (Pluth 2010, Chapter 4). The best – and profoundly damning – critique is Johnston (2008b).
10. In an oft-cited paper Flanagan lists no fewer than fifteen different meanings of the term 'naturalism' and refers to the rejection of supernaturalism as the 'one thing about which all card-carrying naturalists agree, or should agree' (in Clayton and Simpson 2006: 435). Elsewhere, however, Stroud ironically notes that '[i]n the sense in which naturalism is opposed to supernaturalism, there has been no recent naturalistic turn in philosophy. Most philosophers for at least one hundred years have been naturalists in the nonsupernaturalist sense' (in De Caro and Macarthur 2004: 23).
11. Inquiries into the multifaceted, geographically various and at times contradictory concept and history of the 'Enlightenment' easily comprise one of the most burgeoning fields in the whole of contemporary humanistic academia. Skimming the cream of a vast number of historical monographs on the subject, the work of Peter Gay in the late 1960s (Gay 1966, 1969) and that of Jonathan Israel in the early 2000s (Israel 2001, 2006, 2011) are mandatory mentions (and hefty reads). My own views – perhaps biased by contemporaneity – are particularly influenced by the latter, welcoming his warning that 'there are sound, even rather urgent, reasons for rejecting [the Enlightenment critics' arguments] as profoundly misconceived and insisting, on the contrary, that the Enlightenment has been and remains by far the most positive factor shaping contemporary reality and those strands of 'modernity' anyone wishing to live in accord with reason would want to support and contribute to' (2006: v).

12. See Brandom (2002) for an insightful analysis of Kant's normative turn.
13. In the early, and perhaps still best, 1798 translation by John Richardson of Kant's semantically rich *'Aufklärung ist der Ausgang des Menschen aus seiner selbstverschuldeten Unmündigkeit'*.
14. This emerges clearly in Brandom's (German idealism-inspired) social-pragmatist account of knowledge, where – far from a naively Baconian equation of knowledge with power – the attribution of knowledge to a subject means that such subject is at once *entitled* and *committed* to a certain claim. A rational agent can count as 'knowing' (and as having the authority that such an epistemic state grants) only to the extent that she binds herself to (assumes responsibility for) the commitive consequences of an assertion and makes herself available to public challenge on (being asked for reasons for) the commitment undertaken. Phrased as a slogan, the knower always *owes* (to the collectivity of rational agents) as much as she *owns*.
15. The realism I defend in this book parts ways with the transcendental idealist when this assumption of responsibility in the normative and epistemic sphere is extended to the ontological one, making the human responsible for the *constitution* of the real. However, Kant's transition from a theocentric to an anthropocentric paradigm (terms introduced by Henry Allison 2004) and from an intellectualist to a discursive model of cognition is to be recognised as a momentous philosophical achievement, opening the way to the *intrinsically anti-theological* rejection of (in Sellarsian language) the 'myth of the categorial Given' or the belief in the direct sensory awareness of a propositionally articulated reality. (This thesis, regrettably, still enjoys some popularity among theologians and philosophers of religion. Alvin Plantinga, for example, argues that '[f]or science to be successful... there must be a match between our cognitive faculties and the world' (2011: 270) a match – if I can be forgiven the pun – clearly *made in heaven*). The anti-realist outcome of Kantian transcendental idealism can be avoided through a naturalised revision of Kantianism, preserving the ontological dependence of the conceptual on the real. For a reading of Sellars as offering precisely a naturalised Kantianism see O'Shea (2011).
16. The motley crew of counter-Enlightenment thinkers from Hamann and Jacobi all the way to Heidegger, thinkers keen on the Romanticist substitution of universalised 'reason' with contextualised and relativist notions of incommensurable 'traditions', personal 'feeling' or unanalysable 'faith'. See Berlin (1980) for a classic exposition of counter-Enlightenment thinkers, Beiser (1987) for a detailed historical reconstruction of the philosophical and cultural *milieu* in which these ideas first flourished, and Lloyd (2013) for a counter-history of the Enlightenment mindful of its internal contradictions and 'shadows'.
17. As Derrida put it, the ideal of Enlightenment critique, the *krinein* that entails responsibility, can surely be object of deconstructive thought, yet '[t]o say that... this is deconstructible is not the same thing as to disqualify, to negate, deny, or exceed, to perform the *critique of the critique* (just like critiques of the Kantian critique were written as soon as it appeared), but to think its possibility from another point, that of the genealogy of judgment' (1995: 286).
18. Again, Richardson, alone among translators to follow, insightfully translated *Aufklärung* as 'Enlighten*ing*'.

19. Not all Enlightenment thinkers had such an anti-theological agenda. Just consider an influential *Aufklärer* like Lessing who keenly sought to preserve a rational core of Christianity, or indeed Kant's own notion of *Vernunftglaube*. But as I said, that of Enlighten*ing* is a still-active process.
20. I will further comment on the notion of an order of nature in Chapter 3. The important point to insist on here is how a notion of natural order has all too often mandated the extension of its alleged laws to the aesthetic, moral and social spheres. For perspectives on the multifarious ways in which 'nature' has historically been taken to hold authority over the normative sphere see Daston and Vidal (2004) and Watkins (2013).
21. Unqualified endorsements of 'secular reason' all too often fail to account for the complexity of the notion of secularism. The so-called 'secularisation thesis', once the consensus among social scientists, has lost popularity in the last decade, its critics highlighting its theoretical and empirical shortcomings (for a snapshot of the current debates see Warner (2010) and Calhoun et al. (2011)). I see a secular *society* and *politics* as a desirable goal (and inextricable from the practice of naturalist-immanentist explanations) but I am not going to argue for it here (but see Gourgouris, who indeed argues that the question of the secular elucidates 'what it means to encounter one's life as a worldly affair and responsibility that rests on no foundation' (2013: xviii) and as such is indissoluble from the issue of social and epistemic autonomy).

1 Badiou's Mathematical Ontology

1. Significantly (as I will show) theologian Frederiek Depoortere (2009: 103) considers these two 'fundamental decisions' as being 'in particular relevant for theologians', since '[they] concern two traditional attributes of God, namely his infinity and his unity'.
2. For the philosophers Badiou has in mind here God works as a condition of possibility for metaphysics. The genitive 'God of Metaphysics' is at once objective and subjective: it is a rational concept whose referent must be the object of metaphysical (ontological) proofs *and* the guarantor of the very possibility of rational demonstration.
3. Cf. the Badiouian God of Metaphysics with the first of the 'two gods' to which Lovejoy refers to in his reconstruction of Western philosophical theology, the 'Absolute of otherworldliness – self-sufficient, out of time, alien to the categories of ordinary human thought and experience, needing no world of lesser beings to supplement or enhance his own eternal self-contained perfection' (2001 [1936]: 315).
4. Cf. Lovejoy's second God, one 'who emphatically was not self-sufficient nor in any philosophical sense, "absolute": one whose essential nature required the existence of other beings, and not of one kind of these only, but of all kinds which could find a place in the descending scale of the possibilities of reality' (2001 [1936]: 315).
5. Note that this is the ontological stance from which Heidegger launches his critique of mathematised science: the mathematical project flattens the world into a pre-established scheme and 'is the anticipation of the essence of things, of bodies; thus the basic blueprint of the structure of every thing and

its relation to every other thing is sketched in advance' (1993: 292). As such, it forecloses the possibility of fundamental ontology as the revealing dis-closure of Being itself. It induces the oblivion of being. It is unsurprising, then, that Badiou will recoil from this understanding of Being 'as endowment and gift, as presence and opening' (2006a: 9) precisely by linking ontology to mathematics. However, I will identify a troublesome connivance between Heidegger and Badiou, on precisely these themes, in Chapter 5.

6. I refer here to Heidegger's infamous prophecy that

    > philosophy will not be able to effect an immediate transformation of the present condition of the world. This is not only true of philosophy, but of all merely human thought and endeavor. Only a god can save us. The sole possibility that is left for us is to prepare a sort of readiness, through thinking and poetizing, for the appearance of the god or for the absence of the god in the time of foundering (*Untergang*); for in the face of the god who is absent, we founder. (2003: 38)

   Badiou comments that this motto buttresses the spurious hope that 'the thinking that poets teach... may uphold at the heart of nihilism the possibility, devoid of any way or means open to utterance, of a resacralization of the Earth' (1999: 51).
7. What Badiou is rejecting here is the poetisation of philosophy rather than poetry per se (cf. Badiou 1999: 67). Much the same as his master, Badiou shares Plato's ambivalence towards poetry – the philosopher who denounced poetry as a dangerous, potentially corrupting practice (in the *Republic*) yet who philosophised via dramatic dialogical set pieces and created the trope of the philosophical myth.
8. As Badiou puts it, the 'most succinct maxim of modern atheism' is the statement that 'truths have no sense' (2008b: 165).
9. Badiou is adamant in the claim that he is 'absolutely an immanentist' (1994: 87), his whole philosophy developing 'under the merciless rule of immanence' (1994: 124). He insists that he and Deleuze shared 'a... conviction as to what it is possible to demand of philosophy today and the central problem it must deal with: namely an *immanent* conceptualization of multiplicity' (2000a: 4).
10. Badiou explains how his (Platonic) adoption of the mathematical as the pure form of thinking is inherited from Lacan, who before him linked the purely formal nature of mathematics to the evacuation of meaning:

    > [n]othing ties me more to Lacan's teaching than his conviction that the ideal of any thinking is that aspect of it which can be universally transmitted outside of sense. In other words, that senselessness [*l'insensé*] is the primordial attribute of the True. What may be called 'Platonism' is the belief that in order to come close to this ideal, it is necessary to mathematize, by hook or by crook. This is opposed by all the doctrinaires of sense or meaning, be they sophists or hermeneuticists – all of them, at bottom, Aristotelians. (2009: 522)

    Cf. Lacan's claim that 'I place the highest value on [mathematical] discourse precisely because it signifies nothing' (2002: 758).
11. For Badiou, philosophy is *necessarily* anti-religious as it 'commences only with a desacralization: it establishes a regime of discourse that is its own inherent and earthly legitimation, [it] requires that the authority of profound

utterance be interrupted by argumentative secularization' (2008b: 36) Cf. Deleuze and Guattari's claim that '[w]henever there is transcendence, vertical Being, imperial State in the sky or on earth, there is religion; and there is Philosophy whenever there is immanence, even if it functions as arena for the agon and rivalry' (1994: 43).

12. The rejection of a static, monumental historicism which obscures the novelty of the present in favour of a philosophy in the service of the past (see Badiou 1999: 113ff.). For a brief statement of Badiou's (positive) relation to history see Badiou in Riera (2005: 256).
13. Watkin argues, correctly in my opinion, that Badiou's employment of axioms is to be justified by a grounding ethico-political commitment. Watkin notes that '[a]xioms do not "prove" anything, least of all God's non-existence. The secularisation of the infinite is the decision that will achieve the 'better' conclusion that God does not exist, not the proof or revelation of any such fact' (2011: 108) and that there is no way immanently (from within Badiou's axiomatics itself) to justify the choice of axioms. Ultimately, then, 'the "Good" for Badiou can be construed as atheism itself' (2011: 109) and Badiou's 'Epicurean's desire to escape the fear of the gods precedes and demands the ontology of multiplicity that Lucretius furnishes in his atomism and that Badiou secures with his axiomatised ZFC set theory' (2011: 110). I agree that 'the necessity of the assumption that there are no limits to infinite multiplicities seems here to be the result of a desire to rest serene in the primacy of multiplicity over the one' (ibid.), and, to a certain extent, I consider this 'atheistic ontological *ataraxia*' (ibid.) a worthy and self-justifying ethical commitment (and Badiou's Lucretian proclivities certainly vindicate this interpretation). What my naturalisation of Badiou will allow, however, is an *a posteriori* justification of Badiou's foundational axioms through scientific, empirical considerations.
14. In general terms, Badiou reacts against post-Heideggerian phenomenology and the 'theological turn' of this movement in the last decades of the twentieth century. See Janicaud (2000) for a well-known critical assessment of this turn.
15. Cf. Ling's analogous diagnosis: 'if [the equation of mathematics with ontology] is unfounded, so too is Badiou's entire philosophy' (in Bartlett and Clemens 2010: 48). Pluth similarly writes that 'there is no argument offered for the claim that the one is not, and there is no argument for why set theory rather than anything else does ontology' (2010: 45). He stresses, however, that the reason behind these choices can only be assessed on the grounds of how they *will* render possible Badiou's theory of truth and of the subject.
16. Badiou considers his axiomatic methodology to be inherited from the Greek (in Badiouian parlance Pythagorean-Platonic) discovery of the matheme (2006a: 125). Such an innovation engendered a new (non-poetic) kind of ontology: '[t]he particular invention of the Greeks is that being is expressible once a decision of thought subtracts it from any instance of presence' (2006a: 126).
17. For a perceptive reconstruction of Badiou's relation to Cantor and his work, see Tho in Bowden and Duffy (2012).
18. The term 'operation' should not invite a reading of the operation *of* some sort of constitutive power (or subject) which transcends the operation itself.

For Badiou, the count-as-one is not in need of heteronomous causation but is an immanent process of the presentation of situations. See Badiou (2004: 170).

19. If ontology is the science of inconsistent multiplicity, *metaphysics* is defined by Badiou (in Heideggerian spirit and lexicon) as 'the *enframing* of Being by the One' (2006c: 34). See also (2006b: 42).
20. Clemens proposes the straightforward equation 'nothing = Being = inconsistent multiplicity' (2005: 104).
21. It is on the crucial choice between preserving an indifferent, inconsistent, pre-structural 'manifold without a predicate other than its own multiplicity' (2006c: 36) and doing away with any such manifold that I will develop my objections to Badiou.
22. Cf. 'Being can only be thought *more geometrico*' (Badiou 2006b: 96) and '[t]he order and connection of ideas is the same as the order and connection of things' (Spinoza 2002: 247; II, P7).
23. Cf. 'What can rationally be said of being qua being…is said – or rather *written* – as pure mathematics' (2006b: 172–3).
24. Badiou hails Heidegger as 'the last universally recognizable philosopher' (2006a: 1).
25. For De Beistegui, Badiou's project 'consists in a genuinely metaphysical attempt to free thought from the double horizon of unicity (whereby it posits itself as ontology) and transcendence (whereby it is theology)' (in Riera 2005: 45).
26. See Eyers (2013) for a thorough reconstruction of the links between French *épistémologie* and Lacanian-Althusserian structuralism, and the informative discussion between Badiou and Michel Serres (in Tho and Bianco 2013) for a Bachelard-inspired analysis of the use of models in the sciences.
27. Badiou seems to acknowledge this when he admits that '[i]f there are "atoms", they are not, as the materialists of antiquity believed, a second principle of being, the one after the void, but compositions of the void itself, ruled by the ideal laws of the multiple whose axiom system is laid out by ontology' (2006a: 58), even though it is debatable whether the atomists actually considered the void as ontologically prior to atoms. This issue is more visible elsewhere, where Badiou invites the reader to 'pose in the style of ancient atomism, that being is pure multiplicity, without ground nor sense' (2002: 66) or interprets Lucretius as 'an author for whom the Void as such is the original principle' (2004: 105). These claims seem mistakenly to invite a straightforward comparison with his own ontology. As Sedley explains '[t]he atomists decided that they could work with a dualistic scheme such as Parmenides had hypothesised and rejected, by holding that not-being was not, after all, a self-refuting concept. Thus the atomistic universe consists, with enviable simplicity, of being and not-being. Of these, being, or "the existent", is equated with body, not-being, or "the non-existent", with void' (1982: 177). Thus if it is true that the atomists – against Parmenides (for whom Being was a homogeneous One) and against Aristotle (for whom absolute non-being was a contradictory notion) – disseminated being in an infinite plurality of atoms, yet these atoms were conceived as *substantial beings* precisely in opposition with the non-being of the void, nowhere near Badiou's understanding of inconsistent multiplicity as a bottomless plurality of multiplicities.

28. Clemens observes that 'the void becomes *the* atom of Being ... [t]here is no longer any absolute duality of "atoms" and "void"' (2005: 105).
29. On Badiou's part this stance is more explicit and vigorously presented in his early essays, particularly his *Marque et Manque: à propos du Zéro* (in Hallward and Peden 2012). Cf. Derrida's employment of 'the Democritian or Epicurian [term] *stoikheion*', as a replacement for 'trace' or 'writing', on account of its 'greatest generality' which 'extends the mark beyond the verbal sign and even beyond human language' (2007: 360).
30. Lucretius explains that 'when the atoms are being drawn downward through the void by their property of weight, *at absolutely unpredictable times and places* they deflect slightly from their straight course, to a degree that could be described as no more than a shift of movement' (2001: 40–1; II:216; emphasis added). Althusser (and Badiou with him) seizes upon this description and highlights the pure contingency at the origin of Epicurean materialism when asking '[w]hat other philosophy has, in the history of philosophy, defended the thesis that *Swerve was originary*, not derived?' (2006: 169).
31. For a comprehensive and unique *mathematical* introduction to Badiou's *Being and Event* see Baki (2014). I owe much of my understanding of the topic to Burhan, who generously shared with me drafts of some of the chapters of his book.
32. Russell observed

    > that a long line of philosophers, from Zeno to M. Bergson, have based much of their metaphysics upon the supposed impossibility of infinite collections. Broadly speaking, the difficulties were stated by Zeno, and nothing material was added until we reach Bolzano ... Intervening attempts to deal with the problem are futile and negligible. The definitive solution of the difficulties is due ... to Georg Cantor. (1996: 169)

    Likewise, Potter explains that, historically, the possibility of an actual infinity becomes a real one only after Cantor's work, whose theories 'exhibited the contradictions others had claimed to derive from the supposition of infinite sets as confusions resulting from the failure to mark the necessary distinctions with sufficient clarity' (2004: 3).
33. He employed diagonalisation in his later 1891 paper. Intuitively, diagonal arguments, acting on an infinite totality of elements, allow for the production of some new element which plays the paradoxical (antinomical) role of being both a member of the totality and demonstrably *not* one of the infinite members of it. For natural numbers, for example, diagonalisation allows the generation of a real number not in the range of the original natural numbers domain but that nonetheless must be recognised as such.
34. Ordinals are numbers that express an order in a succession, as opposed to a size or magnitude. On the other hand, each ordinal's cardinality (size) is equal to the set of all ordinals before it.
35. Hallett writes that Cantor's metaphysical views 'contribute[d] enormously to the shape and development of the key Cantorian concepts, and that therefore there is a direct route from Cantor's metaphysics to the substance and nature of modern set theory' (1984: x). In Dauben's (1990) account, toward the end of his life, largely shunned by the mathematical community and having been denied prestigious academic placements, Cantor

sought intellectual (and spiritual) refuge among theologians. Dauben writes that

> Cantor ... believed in the absolute truth of his set theory because it had been revealed to him ... from God directly. Thus he may have seen himself not only as God's messenger, accurately recording and transmitting the newly revealed theory of the transfinite numbers but as God's ambassador as well. In fact, Cantor did regard his work as a sacred mission. ... He had been given faith in the necessary truth of his mathematical discoveries by virtue of their having come from, and their being part of, God's infinite wisdom. In return, Cantor hoped to aid the Church in correctly understanding the problem of infinity ... [I]t is surprising that his deep religious convictions have received so little attention in discussions of his development of set theory. (1990: 232)

36. Dauben reports that on

> January 22, 1886, Cantor wrote to the cardinal [Johannes Frantzelin] explaining that in addition to differentiating between the infinite in *natura naturans* and in *natura naturata,* he further distinguished between an '*Infinitum aeternum increatum sive Absolutum,*' reserved for God and his attributes, and an '*Infinitum creatum sive Transfinitum,*' evidenced throughout created nature and exemplified in the actually infinite number of objects in the universe. Cantor added that the important difference between absolute *infinitum* and actual *transfinitum* should not be forgotten. (1990: 145)

37. Badiou (2007c: 9) is aware that 'Cantor was very tormented by the idea that with the actual infinite there is something that is incompatible with the Christian religion' and unequivocally concludes that ultimately 'Cantor was mad'. It has been correctly noted (see Purcell 2012) that Badiou's ontology remains faithful to *Zermelo's* post-Cantorian axiomatic rejection of a mathematical totality (especially through the axioms of separation and foundation – see below) rather than Cantor's since the latter never wavered in his commitment to a – divine – absolute infinity capable of reinscribing the entire mathematical universe. Depoortere (2009) and Reynhout (2011) make the same point to underline the arbitrariness of Badiou's exclusion of an absolute infinity. The former argues that 'it is justified to raise the question of whether the axiom of foundation [which rules out the formations of sets requiring an absolute infinity] ... is not an unacceptable limitation of scope of ontology from the very beginning. Maybe there is more to say about being qua being than Badiou's version of set theory allows' (Depoortere 2009: 120–1). As I will make clear below, I think that there is *less* to say about being *qua* being than either he or Badiou would like. See also Phelps (2013) for a critique of Reynhout's and Depoortere's attempts to re-inject theology into Badiou's system, and a different approach to Badiou's crypto-theology.
38. Most notably the paradoxes arising from the concept of a set of all ordinals (known as the Burali–Forti Paradox) or of the set of all cardinals (known as the Cantor Paradox). Probably the most notorious paradox, because most devastating for logical understanding of set theory, is Russell's Paradox (or the Zermelo–Russell Paradox), which showed irresolvable contradictions arising from the concept of self-reference, and which famously shattered Frege's ambitions to found mathematics on the logical principles presented

in his *Grundgesetze der Arithmetik*. Specifically, Russell's paradox states that the notion of a set *x* of all sets which have the property of not belonging to themselves leads to contradictions, since if such a set *x* belongs to itself, then it cannot belong to itself, whilst if it doesn't, then it must.

39. That is to say, all mathematical objects can be viewed as sets and all mathematical theorems can be proved from the axioms of set theory using logical rules. Set theorist Joan Bagaria, however, cautions:

    > writing out a complete proof using the formal language [of set theory] would be extremely laborious, and the result would not only be very long but also virtually impossible to understand. It is important, however, to convince oneself that in principle it can be done. It is the fact that all standard mathematics can be formulated and developed within the axiomatic system of ZFC that makes *metamathematics* possible, that is, the rigorous mathematical study of mathematics itself. (in Gowers et al. 2008: 622)

40. Badiou adopts an *extensional* specification (as opposed to *intensional*) of sets, that is, one that defines the conditions of membership to a set merely by referring to the *actual* members of the set (whatever their nature), rather than in terms of some specific predicate or distinctive feature that characterises its elements.
41. Note that in his rejection of an intuitive understanding of mathematical object in favour of the purely formal axiomatization of ZFC, Badiou is located as far as possible from Husserl's vision of mathematical objects as always originating in a given *Lebenswelt*. Badiou, like Frege, rejects any empiricist or subjectivist grounding of mathematics. Indeed he makes it very clear that 'what first seduced [him] in [his] mathematical education was the non-subjective, the making possible of a capacity to think outside of all intentionality and subjectivity' (2007a: 88). Badiou's aim is precisely to reverse the logical order: '[i]n the rigorous examination of formalization, one can dispose or place the subject, ultimately *as an effect and not as a cause*' (2007a: 89; emphasis added). Cf. Bachelard's diagnosis that 'the mathematical tool affects the craftsman who uses it. *Homo mathematicus* is taking the place of *homo faber*' (1984 [1934]: 57).
42. It can be proven that there can be only one empty set: as empty it lacks any determination, and thus any instance of it is indiscernible from any other. There is only one empty set and all the Ø tokens refer to just one type. Discerning between two sets depends on the identification of at least one member belonging to one set and not to the other. The empty set, by definition, does not have members, thus it cannot differ. However, it is strictly speaking a mistake to claim that the Void is *one*; in proper Badiouian terms it should be said that the Void is a *unique* name (see Badiou 2006a: 90).
43. Cf. Derrida's explanation that 'writing' indexes precisely 'the signifier's capacity to repeat itself by itself, mechanically, without a living soul to sustain or attend it in its repetition, that is to say, without truth's *presenting itself* anywhere' (2004: 114).
44. Note that here Badiou employs 'text' as a synonym of 'situation': intentional or not, here the Derridean congruence resounds clearly. The Void is the aporetic, retrospectively identified 'origin' of any situation just like the trace is the suturing point of any self-present (con-)text with the 'originary'

movement of *différance*. Elsewhere Badiou writes that the void 'names the undecidable of presentation' (2006a: 55). The first assertion of the Void set precedes and undercuts all concepts and all language: does this contradict the Derridean *'il n'y a pas de hors-texte'* (1974: 158)? Not if we read the Derridean 'text' as more closely analogous to Badiou's 'situation' rather than as a narrowly understood language-bound system. There is indeed *nothing* outside of the text/situation.

45. The impossibility of 'directly' describing inconsistent multiplicity doesn't entail the unspeakability of negative theology. Whereas in the latter the Being of God is so removed from creation that human concepts necessarily fail to grasp it, in the former this impossibility is simply dictated by the total lack of any predicable determination.
46. Throughout this section, I will refer to a series of seminars on the historical development of the concept of infinity delivered by Badiou at the European Graduate School in 2011, available on the EGS YouTube channel.
47. The founding authority on the matter, whose influence would extend from Antiquity onwards (through the influential mediation of Aquinas) was of course Aristotle, who, in his *Physics,* struggled with the paradoxes posed by the concept of infinity, necessary to invoke in the context of his inquiry on time, change and division of magnitudes. The Stagirite then drew a distinction – applying the potential/actual dichotomy which he introduced in the doctrine of hylomorphism in his *Metaphysics* – which stood for centuries to come: between potential infinity and actual infinity (intuitively, the difference between infinity as a *process* and infinity as a *state* or, better, *cognitive object*). Consistently with the empiricist concerns that guide his *Physics,* Aristotle argued that everything that is is infinitely divisible *in potentia,* and yet there is no possible infinity *in actu,* to be *reached* by eternal addition or subtraction (see *Phys.* III.6). For Aristotle, who drew definite lines between metaphysics, physics and mathematics, the role of the physicist is to examine real entities, and the quantitative abstractions of the geometer and the arithmetician cannot be applied to this inquiry into physical reality. Aristotle's universe is a finite one, and 'it is impossible to exceed every definite magnitude, for if it were possible there would be something bigger than the heavens' (*Phys.* III.7, 207b).
48. This is the sequence which, in its 'middle' part, sees the perfecting of the idea of the God of metaphysics (see Section 1.2 above), now essentially identified with the concept of infinity, in the thought of rationalist philosophers. Residual Aristotelian scruples, however, kept these metaphysicians from fully accepting the idea of actual infinity, even regarding God. This tension is well expressed in Descartes' reflections on God's infinity. He observes, in a famous passage from his *Meditations,* that 'God ... I take to be actually infinite, so that nothing can be added to his perfection' (1996: 32). Such an infinity, though, can be inferred as a metaphysical necessity through the use of the *lumen naturalis rationis,* but never fully understood or comprehended since 'to comprehend is to embrace in thought; but to know a thing it is sufficient to touch it with our thought' (2000: 30). Actual infinity remains beyond full conceptual grasp since a finite intellect cannot hope to comprehend an infinite entity: Descartes is aware of this fallacy of reasoning and indeed writes that 'I have tried with care to avoid it, for I have never treated the infinite

except to submit myself to it, and not in the least determine what it is and what it is not' (2000: 95). Badiou's moral is that, until the Cantorian revolution, the infinite remains a limit concept, which can only tentatively be predicated of a transcendent God.

49. Hegel's philosophy of the Absolute thus delineated a new kind of bipartition of infinity: spurious infinity and true infinity. The spurious infinite (*die schlechte Unendlichkeit*), for Hegel, derives from the repeated mutual determination of a series of discrete finite terms: such an infinite process of negation of the finite by another finite engenders a 'bad' (or negative) infinite inasmuch as it is still dependent upon the finite for its development. Hegel (2010: 113) writes:

    > [t]his bad infinite is in itself the same as the perpetual ought; it is indeed the negation of the finite, but in truth it is unable to free itself from it; the finite constantly resurfaces in it as its other, since this infinite only is with reference to the finite, which is its other. The progress to infinity is therefore only repetitious monotony, the one and the same tedious alternation of this finite and infinite.

    On the contrary, true or affirmative infinity is when finite elements are understood as self-sublated into the infinite, when the infinite is a continuous unity of finite and infinite (in Badiou's terms preserving 'the ontological immanence of the one' (2006a: 165)). Thus,

    > [a]ccording to this concept, the unity of the finite and the infinite is not an external bringing together of them, nor an incongruous combination that goes against their nature, one in which inherently separate and opposed terms that exist independently and are consequently incompatible, would be knotted together. Rather, each is itself this unity, and this only as a sublating of itself in which neither would have an advantage over the other in in-itselfness and affirmative existence. (Hegel 2010: 116)

    Badiou will highlight that, for Hegel, bad infinity is essentially the *mathematical* infinite: Hegel, and the whole romantic tradition with him, institutes – on the ground of infinity – a historical separation of philosophy from (and rivalry with) mathematics thanks to his decision that

    > philosophy dealt with *the same thing* as mathematics. The Romantic gesture is based on an identification, not a differentiation. In the realm of the concept of the infinite, Hegelian philosophy claims to constitute a superior mathematics, which is to say, a mathematics that has sublated, overtaken, or left behind its own restricted mathematicity and produced the ultimate philosopheme of its concept ... what is at stake with the complete disjunction of philosophy and mathematics carried out by the Romantic gesture is *the localization of the infinite*. (Badiou 2006b: 37–8)

    In disparaging mathematics, in other words, Hegel reinscribes the infinite within the totalising presence of the Absolute, philosophical thought being its highest self-expression – an infinite whose sublating power can find no match.

50. That is, it was anti-Aristotelian to a large extent *because* it was mathematised. Stephen Gaukroger explains that, in the first half of the seventeenth century, '[t]he single greatest obstacle from the methodological point of view was the objection that mechanics dealt only with mathematical idealizations, not with reality [as matter theory did], and this objection had to be cleared out

of the way before the project for a quantitative natural philosophy could proceed' (in Rutherford 2006: 71). Beyond the rhetorical excesses of scientific hagiography (into which Badiou tends occasionally to slip) Galileo's greatest intellectual achievement was precisely the overcoming of the Aristotelian barrier between, on the one hand, a 'higher' natural philosophy seeking (qualitative) explanations in terms of fundamental constituents (essential properties) of matter and, on the other, the Archimedean tradition of mechanics, employing pragmatically abstract, hybrid mathematical-physical descriptions to account for the behaviour of 'machines' (levers, pulleys, pendulums, and so on.). See also Gaukroger (2006: Chapter 11).

51. Galileo, however, could not accept the idea of a *hierarchy* of infinite, numerical, quantities – something that only the third, Cantorian sequence makes possible. Galileo's intuitions, voiced through his spokesperson Salviati in the *Dialogues Concerning Two New Sciences*, led him to believe that there could not be any meaningful sense in which different sizes (cardinalities) of infinity could be compared:

    > we can only infer that the totality of all numbers is infinite, that the number of squares is infinite, and that the number of their roots is infinite; neither is the number of squares less than the totality of all numbers, nor the latter greater than the former; and finally the attributes 'equal,' 'greater,' and 'less,' are not applicable to infinite, but only to finite, quantities. (1954: 32–3)

52. Badiou here refers to Alexandre Koyré's well-known thesis (in Koyré 1957) of a historical passage in modernity from the closed world to the infinite universe, yet denies that the Galileian conquest was enough, short of the full Cantorian accomplishment of a conceptually tame notion of infinity.
53. Note, however, that Cantor's work did not once and for all resolve the issue of the existence of an actual infinity, since his results were received with some scepticism. Even after Cantor's work, Henri Poincaré would still resolutely claim that 'actual infinity does not exist. What we call infinite is only the endless possibility of creating new objects no matter how many objects exist already' (quoted in Kline 1980: 233).
54. See, for example, Caputo and Vattimo (2007).
55. See, for example, Ursula Goodenough's 'religious naturalism'. Goodenough warns against the 'nihilism' produced by the 'scientific version of how things are' and argues that '[i]f religious emotions can be elicited by natural reality...then the story of Nature has the potential to serve as the cosmos for the global ethos that we need to articulate' if 'we all experience a solemn gratitude that we exist at all, share a reverence for how life works, and acknowledge a deep and complex imperative that life continue', and finally recommends a 'covenant with Mystery' (1998: xvii, 167).
56. This is a philosophy which thus reintroduces the One in the form of a univocity of Being. This is the most damning objection that Badiou raises against Deleuze's project, one where 'contrary to the commonly accepted image (Deleuze as liberating the anarchic multiple of desires and errant drifts), contrary even to the apparent indications of his work that play on the opposition multiple/multiplicities,... it is the occurrence of the One – renamed by Deleuze the One-All – that forms the supreme destination of thought and to which thought is destined' (2000a: 11). Peter Hallward offers a heterodox

interpretation of Deleuze along similar (Badiouian) lines, claiming that 'the logic of Deleuze's work tends to proceed broadly in line with a *theophanic* conception of things, whereby every individual process or thing is conceived as a manifestation or expression of God or a conceptual equivalent of God (pure creative potential, force, energy, life...)' (2006: 4). For another take on the Badiou-Deleuze relationship (and a defence of the latter against the former's accusations) see Roffe (2012).

57. His conception here is opposed to that of Spinoza – the thinker of a *Deus sive Natura* as 'a being absolutely infinite, i.e., a substance consisting of an infinity of attributes, of which each one expresses an eternal and infinite essence' (2002: 217) and thus author of 'the philosophy *par excellence* which *forecloses the void*' (2006a: 113). Spinoza's limit, Badiou argues, is to have eradicated the generative excess of the void by positing a supreme count-as-one. (Cf. Hegel's similar critique of Spinoza's system, where 'everything goes into the abyss but nothing emerges from it' (1990: 155)). On the other hand Badiou defends Spinoza against the Deleuzian interpretation, inviting the rejection of 'all interpretations of Spinoza based on potency in terms of virtuality, on action in terms of actualization, or on desire in terms of creativity of life' since '[i]n the geometrical order, which expresses the divine geometry or the mathematics of Being, nothing is virtual, and everything is actual' (in Vardoulakis 2011: 48).

58. The rational thinkability of the infinite is, for Badiou, a crucial, epochal achievement of human intellect. It is no surprise, then, that he would see any 're-enchantment' of the concept of infinity as a dangerously reactive move. The best (or worst) example of this return to a divine infinity is, undoubtedly, Emmanuel Levinas. Levinas appropriated the idea of infinity and gave it an ethical declension by constructing his philosophy around the concept of the infinite alterity between the subject and the Other: for the subject to enter into relationship with the other without divesting it of its alterity an inexhaustible distance must lie beyond the face of the other, an infinity constitutive of any relationship, making the Other impossible to subsume under a totality of phenomenological presence. For Levinas, then, the infinite indexes the transcendent, the unobjectifiable wholly other, and the divine beyond any ethical encounter. Reflecting on the Cartesian position that the infinite is a rational idea logically prior to the finite, Levinas (1991: 48–9) writes that

    > [t]o be sure, things, mathematical and moral notions are also, according to Descartes, presented to us through their ideas, and are distinct from them. But the idea of infinity is exceptional in that its *ideatum* surpasses its idea, whereas for the things the total coincidence of their 'objective' and 'formal' realities is not precluded...it is of importance to emphasize that the transcendence of the Infinite with respect to the I which is separated from it and which thinks it measures (so to speak) its very infinitude....The transcendent...is infinitely removed from its idea, that is, exterior, because it is infinite. To think the infinite, the transcendent, the Stranger, is hence not to think an object.

59. Hallward is correct when he warns against an identification of inconsistent multiplicity as a post-phenomenological 'Other' since '[t]he "experience" of inconsistency is precisely not an experience, but merely an implication

occasioned by a momentary suspension in the rules that usually make experience intelligible' (2005: 12). However, in my view even this implicative understanding of inconsistent multiplicity does not fully satisfy the empirically minded philosopher. Moreover, the problem remains that as long as there obtains, between the Void and inconsistent multiplicity, a quasi-relationship of sign and signatum, the metaphysical door remains open for the theologian. As Derrida warned '[t]he age of the sign is essentially theological' (1974: 14). Note that Badiou is aware of the existence of a 'religious co-opting' of his work but, significantly, he restricts this possibility to his theory of the event (see Chapter 5): 'when your work concerns the relation between truth and an event you are necessarily exposed to a religious interpretation. You cannot avoid it. You are exposed because you are no longer confined to the strictly empirical or ontological field' (Badiou 2005a: 41). Here I would note how Badiou seems – mistakenly – to consider his ontology as inherently immune to theological reinterpretation.

2 The Ontological and the Empirical: Naturalist Objections

1. I borrow this very useful term, here and in what follows, from Brassier (2007: 114).
2. Opening his extensive study of early Pythagoreanism, Leonid Zhmud claims unequivocally that '[t]he Pythagorean question remains one of the most intricate in the history of early Greek science, philosophy and religion and has every chance of being consigned to the category of insoluble problems' (2012: 1). For another nuanced approach to Pythagoreanism as a contested historiographical category see Cornelli (2013).
3. It is telling that Badiou's example of mathematised physics is the Galilean principle of inertia: the mathematisation of nature has progressed far beyond Galileo in the last five centuries.
4. Badiou himself symptomatically acknowledges the conceptual contortions occurring when he tries to think about the relationship between mathematics and science confessing that that 'the difficulty in my conception is that ontology has to explain why science operates but ontology is mathematics, so mathematics has to explain how mathematics operates and it is a real problem, a real problem…' (2004: 185).
5. I owe the tripartite structuration of the historical evolution of mathematisation that I employ here to Boniolo and Budinich in Boniolo et al. (2005).
6. See Badiou (2009).
7. See Hacking (2000).
8. In her 1997, 2007 and 2011. For an exposition and a critique of naturalism in the philosophy of mathematics (and a defence of Platonism) see Brown (2012).
9. Mathematicians are 'ontologists without knowing so' (2006a: 13). Paraphrasing Althusser (one of Badiou's masters) Badiou implies the spontaneous ontology of the mathematicians.
10. Brassier (in Bartlett and Clemens 2010: 71) outlines a similar contrast, along naturalist lines, between Badiou and Quine.

11. The label 'Platonism' for this sort of mathematical realism is generally traced back to Paul Bernays, who defined it as 'the tendency [which] consists in viewing the objects [of mathematics] as cut off from all links with the reflecting subject' (Bernays in Benacerraf and Putnam 1983: 259). A classical definition of mathematical Platonism is G.H. Hardy's: 'I believe that mathematical reality lies outside us, that our function is to discover or observe it, and that the theorems which we prove, and which we describe grandiloquently as our 'creations', are simply our notes of our observations' (2005: 35). More operationally, a Platonist treats the objects of a given mathematical theory as if they were independent of what he or she can explicitly define. See Panza and Sereni (2013) for a discussion of Platonism's irresolved conceptual problems.
12. The concept of *arithmos* employed by Greek mathematicians (and known by Plato) was radically different to the post-Renaissance understanding of number which we have today. In Greek mathematics, *arithmos* indicates a finite collection (or set) of items, and arithmetic was the study of the ratios of magnitude obtaining between these collections (consider how the English 'calculus' derives from the Latin *calculus* meaning 'pebble', those small stones moved around an abacus for, precisely, calculations). It is only after the fifteenth century that the notion of number as an abstract entity becomes possible (thanks in part to the introduction in Europe of Indo-Arabic numerals). As Pritchard explains, it is post-Renaissance mathematics that

    > now deals with things which are not, strictly speaking, imaginable – not in the sense that something infinite is unimaginable, but rather because mathematics now no longer deals with objects which are given in a certain way (as lines, surfaces, solids, times, *arithmoi* etc.), but treats directly of the relations themselves, which hold between things which can be of any kind whatsoever... [T]he [Greek] mathematicians pictured *arithmoi* as sets of points because *arithmoi* are sets of units. Number (as 'abstracted ratio') cannot be strictly pictured at all. The abstraction required to reach number is a higher level of abstraction which does not use the imagination and is utterly alien to Greek notions... [T]he so-called 'Platonist" philosophy (or philosophies) of (modern) mathematics owe nothing to Plato except for the spurious respectability derived from attaching his name to a set of views which he never held, and of which, could he understand them, he would be unlikely to approve. (1995: 44–5, 46, 177)

    Pritchard's account of Greek mathematics corroborates Badiou's claim that among the Greeks 'the being of number is the multiple reduced to the pure combinatorial legislation of the one' (2006b: 61).
13. Badiou's Platonism is a specifically *French* kind of Platonism. He explains that, differently from Gödel's 'Platonism of ideal objects' developed within and against the framework of 'American empiricism', for French philosophers 'the question of dialectics has always been central' and 'the heart of dialectics in Plato is the question of participation. It is not so much the distinction between the sensible and the intelligible' (2007a: 93). Cf. Lautman's (one of Badiou's mathematical masters) 'dialectical platonism', opposing an (Anglo-American) notion of Platonism as entailing the mind-independent existence of mathematical objects, product of 'a superficial knowledge of Platonism' (2011: 190). Zalamea similarly argues that Badiou's ontology 'allows us to bring an *untrivialized Plato* back to the landscape of mathematical philosophy'

(2012: 279). On Badiou's relation to Lautman see Bowden in Bowden and Duffy (2012).

14. I disagree with Gillespie's claim that Badiou's position is a kind of formalism. If it is certainly true that for Badiou 'mathematics is concerned with the formal groupings and ordering of multiplicities rather than with any logical relations between a set and its elements' (Gillespie 2008: 50) this does not suffice to equate the two stances. If the statement 'mathematical signs refer to nothing' could be attributed to both Badiou and Hilbert the 'nothing' in question is *not* the same (recall that inconsistent multiplicity is no-thing, not a nothing). Hilbert's formalism conceiving mathematics as the meaningless manipulation of symbols leaves no space for the kind of ontological value that Badiou wants to assign to mathematics: suffice it to say that Hilbert's neo-Kantian rejection of any use for the concept of infinity (beyond that of a merely regulative idea – see Hilbert in Benacerraf and Putnam 1983: 201) is an outcome of his formalist project profoundly at odds with Badiou's own ontological commitments.
15. One could retort that these concerns are today somewhat dated, grounded on an early twentieth-century concern with foundationalist projects, and out of touch with the contemporary state of mathematical practice. Zalamea (2012) mounts a powerful argument precisely to this effect. I have, unfortunately, come around to read Zalamea's work only late in the writing process of this book, and I cannot engage with his powerful approach to the philosophy of mathematics in the depth it deserves.
16. Shapiro writes that 'any philosophy of mathematics or philosophy of science that does not provide an account of this relationship [between mathematics and the rest of scientific discourse] is incomplete at best' (2000: 34). I will examine Shapiro's own structuralist answer to this problem in Chapter 4.
17. Wilholt (2006) has convincingly argued that the problem of applicability disappeared for a long time from the philosopher's agenda because of the influence of logical positivism. The post-Fregean adherence to the logicist programme allowed the logical positivists to explain away the problem, since their opinion was that all mathematical statements are really grounded on analytic propositions of logic. It was thus no longer a matter of finding a match between mathematical statements and a mathematical (or physical) realm, mathematics being just a highly sophisticated system of tautologies. This thesis has been developed further by Ginammi (2011). Steiner summarises this stance thus: '[h]ow can the abstract entities of mathematics relate to the world of physics? Frege's answer was: they do not; they are related to the laws of the world, not to the world itself' (1998: 47). This position, of course, deprives mathematics of any creative power. A.J. Ayer makes it explicit in a vivid analogy:

    > [t]he power of logic and mathematics to surprise us depends, like their usefulness, on the limitations of our reason. A being whose intellect was infinitely powerful would take no interest in logic and mathematics. For he would be able to see at a glance everything that his definitions implied, and, accordingly, could never learn anything from logical inference which he was not fully conscious of already. (1971: 82)

    In this case, I will concede to Badiou, we can talk of 'little style' (where mathematics is merely a tool for the algorithmic recapitulation of already-present knowledge), but this kind of logicist approach to mathematics has long lost its prominent place.

18. Colyvan (2001b) has shown how Wigner's problem is an issue to be resolved whatever one's metaphysical position regarding mathematical objects happens to be, and he argued that both Field's fictionalism and Quine's indispensability argument are not able to explain it away. Colyvan concludes that the quandaries engendered by Wigner's problem 'seem to cut across the realism/anti-realism debate and thus deserve careful attention from contemporary philosophers of all stripes – realists and anti-realists alike' (2001b: 274). For a disagreement, see Rivadulla in Boniolo et al. (2005).
19. In his 1989 article Steiner talks of 'first-order analogies' (1989: 452) between past experience and future experience.
20. These are also defined as 'second-order mathematical analogies'' (1989: 452).
21. Cf. my discussion in Chapter 4 of John Worrall's employment of the same historical example.
22. Steiner offers, as further examples of this Pythagorean strategy, Erwin Schrödinger's discovery of wave mechanics and Paul Dirac's 'Pythagorean prediction of the positron' (1998: 83).
23. Also defined as 'third-order mathematical analogies' (Steiner 1989: 452).
24. Sarukkai further endorses this analogical role of formalism in mathematical discovery arguing that 'the effectiveness of mathematization significantly depends on the power of symbols to act like pictures of ideas, concepts and events. The role of mathematics in the sciences seems to be essentially dependent on the possibility of using mathematical symbols as "pictures"' (2005: 422). See also Brown (2008) for the role of pictures and diagrams in mathematical proof, and Ippoliti (2008) for an argument for the indispensability of visualisation as a tool for abductive reasoning in mathematics.
25. In his later intervention (in Shapiro 2005) Steiner talks of metaphysical Pythagoreanism – suggested by the omnipresent employment of mathematical formalism to describe fundamental physical entities in contemporary sciences – as 'the ultimate irony' since 'Pythagoras and Democritus might turn out ultimately to have been saying exactly the same thing – of course the world is made only of matter, but look what matter is! Materialism as a doctrine might turn to be not so much wrong as pointless – in a world in which matter, energy and space-time turn out to be mathematical structures' (in Shapiro 2005: 648).
26. In subsequent work Maddy distanced herself from the realist stance she defended in this early book. However, this was not caused by a perceived inadequacy of her own previous position, but by a shift in her methodological priorities. In her 1997 and especially her 2007 monographs Maddy developed a stronger commitment to mathematical naturalism, thus rejecting the *necessity* of offering extra-mathematical justification for the grounds of mathematics (the axioms of set theory): '[w]here Quine takes science to be independent of first philosophy, my naturalist takes mathematics to be independent of both first philosophy and natural science (including the naturalized philosophy that is continuous with science) – in short, from any external standard' (1997: 184). As she explains again in her later book: 'just as a fundamentally naturalistic perspective counts against *criticizing* a bit of mathematics on the basis of extra-mathematical considerations, it counts just as heavily against *supporting* a bit of mathematics on the basis of

extra-mathematical considerations' (2011: ix). I take a more relaxed attitude regarding mathematical naturalism, claiming – with Badiou – that mathematics has an extra-mathematical significance which cannot be assessed purely 'from within' but that requires the development of a complete metaphysical picture of reality.

27. See Goldman's (1967) and Grice (1961) for early statements.
28. However, see Steiner (1973) for an account of causal theory of knowledge-friendly Platonism. Many have chosen the first horn of the dilemma, preferring to give up belief in mathematical objects to preserve the meaningfulness and objectivity of mathematical discourse, thus preserving it as a subject with no object (as Burgess and Rosen (1997) title their survey of these positions). See, most famously, Hartry Field's nominalist (or fictionalist) project of a 'science without numbers' in Field (1980). A more recent defence of nominalism (under the name of the 'constructibility thesis') is Chihara (1990). See Kitcher (2012) for a recent deflation of Benacerraf's dilemma through a historicist modification (or pragmatist abandonment) of the idea of Tarskian truth in mathematics.
29. See Quine (1953) and Putnam (1971) for classic formulations, Colyvan (2001a), Resnik in Shapiro (2005) and Panza and Sereni 2013 (Chapters 6 and 7) for critical surveys of the argument, and Bangu (2012) for a recent re-elaboration and defence.
30. This of course does away with the idea of 'abstract object'. Maddy doesn't seem to mind: 'so be it; I attach no importance to the term' (1990: 59). Wherever some physical object which instantiates the set lies, there the set can be found. This means as well that Maddy is happy to do away with so-called 'pure sets', *including the empty set,* since 'each set, no matter how exalted in rank, is located where the physical stuff in its transitive closure is located' (1990: 156). There seems to be here an irreconcilable disagreement with Badiou, whose ontological reliance on the empty set I have explained above. Not so, however, if we accept Maddy's account as an *epistemological* account of how we come to understand numbers: I will argue that such an account need not exclude a physics-informed matherialism 'founded' on the empty set. Whatever the empty set is, it is quite clear that it cannot be an object of the kind of sensible perception Maddy is referring to here.
31. Maddy indeed recognises that her position, depending on physical instantiations of sets, could be characterised as more Aristotelian than Platonic but says that she 'retain[s] the term "Platonism" here, not for its allusion to Plato, but because it has become standard in the philosophy of mathematics for any position that includes the objective existence of mathematical entities' (1990: 158).
32. These are synapses where the input neuron's stimulation grows in efficiency the more it is activated. When a number of neurons habitually firing together in presence of a stimulus strengthen their synapses they become able to activate each other in the absence of the stimulus, and thus compose what Hebb called a 'cell assembly', a network of neurons creating an internal representation of the stimulus. Hebb's theory introduced the concept of a permanent change induced by environmental stimuli and that of brain *plasticity,* and naturalised the very concept of 'ideas' by offering a physicalist explanation for their creation and substistence. See Hebb (2002). For an assessment of

Hebb's work and its influence on contemporary cognitive psychology see Milner in Levitin (2002) and Brown and Milner (2003).

33. Johnston's critique of Badiou's ontology is grounded precisely on the observation that Badiou ignores the 'Hebb-event', or how Hebb showed that 'experiential, pattern-based associative activations of multitudes of neurons in the brain actually alter the calibration of the strength of synaptic connections between these neurons, thus resulting in physical changes in the folded matter of the nervous system' (Johnston 2008a: 41), i.e. what has since been referred to as 'neuroplasticity'. Like Johnston, I believe that a truly atheistic materialism will reject any form of dualism, whether between matter and spirit or between physical and mathematical reality. Unlike Johnston, I think that an account of the immanent origin of mathematical concepts from the neural plasticity of the brain does not exclude the possibility that, at an elementary level, reality (the very same reality to which our brains belong) is a mathematically describable structure.
34. See Brannon and Merritt in Dehaene and Brannon (2011).
35. For a thorough defence of this thesis, with particular emphasis on the role of conceptual metaphors in mathematical thought see Lakoff and Núñez (2000).
36. An even more byzantine subdivision of reality into separate realms is endorsed by Roger Penrose (2004, Chapter 1 and in Polkinghorne 2011), who postulates the existence of three 'worlds' (the mathematical, the physical and the mental) bringing about the necessity of accounting for the three 'deeply mysterious connections between them' (in Polkinghorne 2011: 41). A relevant debate on possible interactions amongst the three realm is Hut et al. (2006).
37. Cf. the so-called 'Extended Mind' thesis, which rejects the restriction of 'mind' to our biological brain and emphasises the role that representational tools in the environment play in cognition, to the point of considering them as part of 'mind' itself. The standard references are Clark (2008) and the collection of essays in Menary (2010).
38. See Zhang and Norman (1994) for an influential analysis of 'distributed cognitive tasks' in scientific and everyday actions, emphasising the role of *external representations* alongside internal ones: physical objects and symbols in the environment or relations between them, in the form of diagrams or abacuses.
39. The practice of carving tallying sticks 'probably dates back at least to the Upper Paleolithic period (35,000–10,000 bc), when anatomically modern humans notched bones and possibly other perishable materials' (Chrisomalis in Robson and Stedall 2009: 503).
40. Derrida, more than Badiou (whose rarefied understanding of mathematics as a set of pure, rational laws of thought does not concede anything to empirical genesis) was very clear about the logical and metaphysical priority of material inscription. In his early work on Husserl (Derrida 1989) he observed how the latter's regressive analysis of the objectivity of geometrical ideal objects depends on the diacritical writing (as differential system of marks) which allows for their repetition (iterability) through a series of virtualisations and actualisations. Derrida's concept of arche-*writing* suggests precisely that inscription, pro-gramming (physical as well as biological, or even

cybernetical (Derrida 1974: 9)) is the condition of possibility for any self-presence, any ideality, and any life. Writing is not a (secondary, dispensable) supplement to the alleged immediacy of mathematical thought and the lifeless (material) substrate is indeed necessary for any signification. '*[T]heoretical* mathematics', argued Derrida, 'whether understood as sensible *graphie*...,or understood as the ideal synthesis of signifieds or a trace operative on another level, or whether it is understood, more profoundly, as the *passage* of the one to the other – has never been absolutely linked with a phonetic production' (1974: 9).

41. See Polkinghorne (2011).
42. Badiou's opinions on the historical evolution of mathematics are somewhat *ad hoc*. In this regard, Roffe has correctly identified an inconsistency in Badiou's prioritisation of set theory: 'if mathematics has always been the expression of being *qua* being, then its historical development, only realized in the advent of set theory, must be merely secondary. But Badiou's claims about set theory seem absolutely primary. In either case, the relationship that exists between being and thought in mathematics seems traversed by a problematic historicity in Badiou's presentation' (2006: 332).
43. Notice that this second, metaphysical claim corrects the potentially relativistic potential of the first, epistemological one. That is to say, unlike scholars who have tried to argue against the ideal of mathematics as *the* form of objective and infallible deductive knowledge by revealing the sociocultural origins of its practice (for constructivist and conventionalist variations on this theme see, for example, Bloor 1976, Porter 1995, Ernest 1998 and Wittgenstein 1976), I claim that while the origins of mathematical thought are to be historically contextualised and biologically constrained, *still* a reasonable claim can be made that the kind of mathematical structures articulated through mathematical thought, and employed in the description of subatomic physical reality, enjoy a mind-independent status.
44. Hallward comments that 'from the outset, Badiou banishes from ontology precisely this general "feel for the world" as no more than an invitation to sensual and ideological confusion. The question is whether the resulting clarity can ever adequately move beyond its operational abstraction' (2004: 106). It is obvious that my insistence on reconstructing a link (or better, erasing the difference) between mathematical abstraction and physical concreteness, is no concession to phenomenology.
45. 'Quasi-' because of Badiou's ostensible rejection of Heidegger's ultimate relapsing into sacralised presence (and passively received truth) but also because, as I have observed above, of Badiou's unwitting multiplication of the problem of ontological difference, summarised in the triad of inconsistent multiplicity, consistent/presented multiples *and* natural multiples/physical reality. One could arguably find for a more elegant (and *naturalisable*) 'resolution' of the problematic of ontological difference in Derrida's pushing of the Heideggerian project beyond Heidegger himself, affirming that 'within the decisive concept of ontico-ontological difference, *all is not to be thought at one go*; entity and being, ontic and ontological, "ontico-ontological," are, in an original style, *derivative* with regard to difference; and with respect to what I shall later call *différance*... The ontico-ontological difference and its ground (*Grund*) in the "transcendence of Dasein"...are not absolutely originary.

Différance by itself would be more "originary"' (1974: 23). See my discussion of Luciano Floridi's defence of structural realism in Section 4.3 for a similar vouching for the precedence of difference.

46. I see this as a *demystification* of Badiou's system. Depoortere makes a valid point when he notes that 'for Cantor inconsistency comes at the end, so to speak, at the limit of his mathematical endeavours. There, where the count-as-one fails, one bumps into the absolute. For Badiou, in contrast, inconsistency is primary: it is the nothing that precedes the count-as-one. So, while Badiou adopts Cantor's distinction between consistent and inconsistent multiplicities, he turns the distinction upside down' (2009: 114–15). Even observing Badiou's warnings against the 'Great Temptation' (of interpreting this Being as a transcendent and inexpressible presence beyond all presented beings) the danger of a theological reappropriation and, more generally, the question of the very intelligibility (from a materialist standpoint) of the presented/pre-presented duality remain as murky shadows over Badiou's purportedly limpid system.
47. Giuseppe Longo offers a recent defence of the thesis that

    > the objects of mathematics are 'outside of ourselves' (transcendent) only as much as they belong to a constituted, which precedes our subject: they are a co-constituted, at the same time as the very intelligibility of the world, by our 'living and communicating community.' They are not arbitrary because they are rooted in the regularities of reality, to which are confronted our living beings in the world. (Longo in Longo and Bailey 2011: 29)

    Herman Weyl and L.E.J. Brouwer are both pre-eminent forebears in this Husserlian tradition.
48. But see also J.J. Katz's defence of a full-blooded 'realistic rationalism' and his claim, against Benecerraf-style epistemological objections that '[mathematical] [r]ealism without rationalism is unbelievable and rationalism without realism is unstable...The integration of realism and rationalism in a single position provides realism with epistemological credibility and rationalism with ontological stability' (Katz 1998: xxxii).
49. Perhaps I should add a fifth non-answer to the problem, a deflationary or downright pessimistic stance forgoing the possibility of ever offering an explanation for the applicability of mathematics. This seems often to be the stance of non-philosophers (mathematicians or natural scientists). An eminent example is Bourbaki, claiming that 'there is an intimate connection between experimental phenomena and mathematical structures [which] seems to be fully confirmed in the most unexpected manner by the recent discoveries of contemporary physics' but adding that 'we are completely ignorant as to the underlying reasons for this fact (supposing that one could indeed attribute a meaning to these words) and we shall perhaps always remain ignorant of them' (1950: 231).
50. This is, of course, the centuries-, if not millennia-, old belief in an ordered cosmos as a 'coherent, luminous, intellectually secure and dependable world, in which the mind of [humans] could go about its business of seeking an understanding of things in full confidence' (Lovejoy 2001 [1936]: 328).
51. I think here of Heideggerian *Gelassenheit* and its post-Heideggerian, phenomenological permutations.

3 Taking a Stance on Realism and Naturalism

1. This is, of course, a rough taxonomy. A stance must be taken on the ontological polarity between materialism and idealism and that of the epistemological between empiricism and rationalism. For example, according to the criterion of access above, Plato would count as a realist, and one could be a hard-line objectivist without claiming direct epistemic access to the postulated objective reality. Not all 'realism' goes with naturalist and materialist commitments. The scientific realism defended in what follows, however, does.
2. The best contemporary monographs offering a comprehensive treatment of scientific realism are Chakravartty (2007); Kukla (1998); Ladyman and Ross (2007); Leplin (1997); Niiniluoto (1999); Norris (2004); Psillos (1999, 2009); Sankey (2008). Interventions old enough to be now considered 'classics' are (to name only some of the most influential) Bhaskar (2008); Boyd (1980, 1983); Devitt (1997); Maxwell in Feigl and Maxwell (1962); Musgrave in Nola (1988); Putnam (1975); Sellars (1963, 1965); Smart (1963). Finally, good selections of relevant papers for and against scientific realism are anthologised in Curd and Cover (1998); Klee (1998); Leplin (1984); Rosenberg and Balashov (2002).
3. For overviews of the contested notion of 'scientific method' see Kosso (2011) and Nola and Sankey (2007).
4. I adopt this Lockean term somewhat loosely, as for Locke the main primary quality of entities is *solidity* (1996: 42; II. IV. 1). This is today called a *dispositional* property and one which, in a particle physics-informed ontology, cannot be said to be part of the core properties of an entity. The one 'primary quality' admitted by contemporary mathematised physics, grounding a robust conception of objectivity as invariance under transformations, is the structural notion of *symmetry*. See Kosso in Brading and Castellani (2003).
5. Sellars argued that the scientific image can be equivalently re-labelled as the '"postulational" or "theoretical" image' (1963: 7) to the extent that it commits one to explanations of the behaviour of macroscopic entities with reference to their (postulated) microscopic, unobservable constituents – a difference, for Sellars, to be articulated on the epistemic level: concepts indexing microscopic entities 'become observable' by acquiring a non-inferential role in the conceptual scheme of the scientific image (see Brandom in Smith 2002 for a Sellarsian account of observability of subatomic particles). One should perhaps comment on the difference between unobservables *in practice* and unobservables *in principle*. Science shows that there are thresholds of observability; for example, that which is below the limit of optical resolution (depending on the determinate wavelength of observable light) cannot in principle be observed, where 'observed' here means something like 'being object of a perception based on visible-light wavelength stimulus on our perceptual organs'. Different kind of technologies, from electron microscopes to particle detectors, supplement optical limitations by delivering an optical representation of an optically unobservable entity – given that the causal chain leading from the latter to the former – can be explained, these entities (from DNA chains to tau neutrinos) can be said to be observables. However, contemporary physics suggests that at a given scale the extremely

small cannot in principle be the object of *any* kind of representation. The only access to this realm is through mathematical formalism. As Miller has shown, late twentieth-century fundamental physics (from Heisenberg onward) has given up the notion of the visualisability of theories by analogy with observable-world imagery and replaced it with diagrammatic representations (primarily Feynman diagrams) *produced from* mathematical formalisms, so that 'Feynman diagrams *represent* interactions among elementary particles in a realistic manner – that is, there is ontological content to these diagrams' (in Nye 2002: 208). It is thus the case that certain entities are *in principle* unobservable and unrepresentable *except* through the mathematical formalism used to refer to them.

6. These terms are, for Sellars, idealisations describing two distinct but interdependent conceptual frameworks. Seldom mentioned, Sellars talks also of an '"original" image' (1963: 7): the three terms could be described as successively emerging through an evolutionary or ontogenetic processes (Garfield 2012 is particularly insightful here). The original image is the primordial conceptual framework of human beings attributing intentionality to *all* entities, indiscriminately assigning purposive action to fellow humans and objects or natural phenomena (universalising what Dennett 1989 would call the 'intentional stance'). The manifest image is a (jointly empirical and categorial) *refinement* or *sophistication* of the original image, wherein for the first time humans come to recognise themselves alone as *persons* belonging to a community. The concepts of the manifest image thus attempt to make sense of reality on the scale of (and according to the primary concerns of) the individual person as the basic epistemic unit: as such it is *not* an unreflective, naive categorisation of reality, but already bears the marks of scientific inductive method. The *scientific image,* finally, is the most recent and refined conceptual framework humans have developed, which includes a type of reasoning absent from the manifest image, 'namely that which involves the postulation of imperceptible entities, and principles pertaining to them, to explain the behaviour of perceptible things' (1963: 7).
7. See Eddington (1929).
8. This point weighs against the strict distinction between observables and unobservables on which radical empiricist positions like Carnap's or van Fraassen's pivot, one explicitly grounded on the contingent fact of what is observable by human bodies. Van Fraassen explicitly states this point when he argues that '[t]he human organism is, from the point of view of physics, a certain kind of measuring apparatus. As such it has certain inherent limitations – which will be described in detail in the final physics and biology. It is these limitations to which the "able" in "observable" refers – our limitations, *qua* human beings'. However, where are we to draw the observable/unobservable line (an ontological line for the empiricist, deciding on existence or non-existence) when our senses can be augmented by technical apparatuses? Do optical microscopes give us access to observables? Electron microscopes? What about the not so distant future when prosthetic enhancement will allow humans to, say, 'see' electromagnetic radiation beyond the visible spectrum? Grover Maxwell forcefully argues against this epistemic distinction, writing that '[a]lthough there certainly *is* a continuous transition between observability and unobservability, any talk of such continuity from

full-blown existence to nonexistence is, clearly, nonsense' (Maxwell in Curd and Cover 1998: 1057). For further critiques of van Fraassen's distinction between observables and unobservables see Hacking (1981), Churchland in Churchland and Hooker (1985) and Norris (1997, Chapters 6 and 7). See Wilson (1995) for an empiricism-friendly historical analysis of the birth of microscopy and its epistemological consequences.

9. The failure of Kant's epistemology to fit a post-Euclidean and post-Newtonian universe does not force us to give up on the transcendental project altogether: there is still the live possibility of generalising it beyond Kant, through resources that were unavailable to him. For a systematic updating of the transcendental project vis-à-vis contemporary physics see Bitbol et al. (2009), a collection of interventions sharing the common belief that 'adapting transcendentalism [to contemporary physics] is much more fruitful than rejecting it' (2009: 3). For a neo-Kantian defence of the *a priori* in scientific reasoning see Friedman (2001) and Friedman in Boghossian and Peacocke (2000).

10. In Patricia Churchland's memorable formulation,

    > [l]ooked at from an evolutionary point of view, the principal function of nervous systems is to enable the organism to move appropriately. Boiled down to essentials, a nervous system enables the organism to succeed in the four F's: feeding, fleeing, fighting, and reproducing. The principal chore of nervous systems is to get the body parts where they should be in order that the organism may survive. (1987: 548–9)

    This passage is often cited by Alvin Plantinga as a buttress for his argument for the incompatibility of naturalism and evolution, since '[w]hat evolution underwrites is only (at most) that our behaviour is reasonably adaptive to the circumstances in which our ancestors found themselves; hence it does not guarantee mostly true or verisimilitudinous beliefs' including the belief in naturalism (2011: 315). Indeed, *evolution* does not provide reliable beliefs. The belief that the sun circles the earth might be evolutionarily useful but is incorrect. It is *scientific method* that, disengaging thought from the manifest evidence of the senses, provides mostly true or verisimilitudinous (testable) beliefs, and thus supports the naturalist stance.

11. I will have more to say about this in Chapter 5, when I will show how this claim can be demonstrated not to rely on supernaturalist intuitions but, on the contrary, can be made rigorous once my thesis of a mathematically describable structure of reality is accompanied with meta-mathematical reflections on the structurally aporetic nature of mathematics.

12. Psillos's rejection of principled epistemic barriers, vividly worded in the claim that 'the forever secrets of nature (if there are such) do not fall nicely within a conceptual category (the unknown X: the noumena; the non-structure; the intrinsic properties; the auxiliary properties; whatever-there-is-only-thin-epistemic-access-to; whatever-there-is-only-theory-mediated-access-to; and the like)' (2009: xiv) is to be accepted for the straightforward reason that the unknown part of reality (if there is any) is that which cannot *by definition* fall into a conceptual category. This is no neo-Kantian epistemic censure, however, to the extent that our arsenal of concepts can evolve along both the rational and the biological axes. The evolutionary argument undermining our cognitive omnipotence does not exclude (but strengthens our

warrant for) to the possibility of a trans-human future increase in our cognitive capacities, with the subsequent shrinkage of the domain of unknowable phenomena.

13. I reject Putnam's suggestion that the metaphysical realist's (or, equivalently, the externalist's) 'favorite point of view is a God's Eye point of view' (1981: 49), not because I want to endorse Putnam's (post-1979) defence of an inescapably 'internalist' perspective on the external world, but because the metaphor of a God's Eye point of view fails to portray the realist's regulative ideal of progressively accurate description.
14. Against causal theories of explanation, van Fraassen (1980) offers a theory of explanation resting on *pragmatic* theoretical virtues. For an early questioning of his theory's ability to solve problems related to explanation (and avoiding thoroughgoing relativism) see Kitcher and Salmon (1987). Van Fraassen (an empiricist, not an *irrationalist*) concedes that '[c]ommon concern with the rationality of scientific change may take us out of this impasse [the realist demand for explanation *versus* the empiricist rejection of such a demand]' (2006: 288, n. 11). The catch, however, lies in van Fraassen's *empiricist* understanding of rationality, rejecting rule-following as a *metaphysically biased* criterion.
15. In her survey of cognitive psychology's inquiries on the role of explanation, Tania Lombrozo concludes that explanation is an indispensable element of our inferential practices: 'the generation and evaluation of explanations can constrain inferences by appropriately summoning prior beliefs. In evaluating claims, the existence of explanations can constitute evidence, and serve as a basis for eliminating possibilities to assess probability. In generalizing from facts or examples, explanations subsume provided information under a general pattern, thereby highlighting the senses of similarity that warrant induction' and indeed that 'explanation provides a unique window onto the mechanisms of learning and inference' (2006: 468).
16. In his fictional reconstruction of a debate between a constructive empiricist and a (Sellarsian) scientific realist, Gutting has the former object to the latter that '*just like the theist* with his cosmological argument, you make your case by insisting on the need to explain something that there is no reason to think has to be explained' (Gutting in Rosenberg and Balashov 2002: 237; emphasis added).
17. To deny this transcendence, and argue for the inevitable revelation of the transcendent object in the immanence of the subject's experience would amount to (and in fact is the starting point of) Husserlian phenomenology and of the post-Heideggerian French phenomenological tradition as a whole, a philosophical approach that was, as Geroulanos observes, 'marked by an obsession with, and a radical objection to, classical transcendence, by the subject's (lack of) access to transcendence or reliance on it' (2010: 80).
18. This is a current he considers to be a minority in the continental tradition. See Braver (2007) for a comprehensive reconstruction of continental anti-realism.
19. Connected with but distinct from the kind of *epistemic* novelty which is a *desideratum* of any new scientific theory. For a thorough treatment of novel predictive success as a crucial criterion for scientific (realist) epistemology see Leplin (1997).

20. The issues deriving from the Kantian binary of receptivity and spontaneity are vexed epistemological quandaries. The challenge is to embrace receptivity without envisioning it as passive expectancy of a self-disclosing Being (to stick with Heideggerian resonances) and to embrace spontaneity without conceding world-making power to our conceptual schemes. Roy Bhaskar's critical (or transcendental) realism remains a powerful formulation of this compromise position, marrying a commitment to the intransitive dimension of the reality studied by science with attention to the sociological mechanisms operating in the transitive process of theory-formation – and in fact showing that in order to make sense of the latter, one needs a transcendental commitment to the former (see Bhaskar 2008). While I agree with Cazeaux's claim (2007: esp. Chapters 6 and 7) that certain epistemological disputes could be solved (or at least ameliorated) by a shift in the metaphors we employ to describe cognitive access to the world – organised, in Kantian form, around an external–internal matrix – I doubt that the Heideggerian metaphor of a primordial co-implication of subject and object, mutually revealing in an historical process of disclosure or opening, can help us (at least, 'us' as *realists*). If for no other reason this is because, as Braver (2007) has thoroughly demonstrated, the Heideggerian paradigm *has already been applied* in most parts of twentieth-century continental philosophy, invariably inducing an anti-realist attitude grounded on the existential finitude and inescapable historical conditioning of *Dasein*'s cognitive powers (for the closest and most refined approximation to an 'Heideggerian' philosophy of science – one, unsurprisingly, relinquishing realism – see Ginev 2011). While I see it as a grave mistake to fix as one's theoretical objective to go – whatever the price – 'after finitude', as Meillassoux titled his influential 2008 volume (a *faux pas* because the price Meillassoux paid is that of taking a step *backwards* to an hyper-rationalism grounded on intellectual intuition), it is paramount for the realist (as I observed above) to reject post-existentialist finitude and replace it with careful reflection on biologically-imposed cognitive limits, a reflection in line with (and in fact dependent on) reliance on the scientific approximation to truth about the universe.
21. See, for example, how Thomas Nagel's most recent call for a return of teleological explanations in science is construed as a critique of 'the present climate of a dominant scientific naturalism, heavily dependent on speculative Darwinian explanations of practically everything, and armed to the teeth against attacks from religion' and gains support from the (at first sight) disturbingly *reasonable* justification that 'I have thought it useful to speculate about possible alternatives. Above all, I would like to extend the boundaries of what is not regarded as unthinkable, in light of how little we really understand about the world' (Nagel 2012: 127). I want to stress that naturalism can be defended as the only responsible game in the philosophical town *without* construing it as operating reactionary conceptual policing on the boundaries of the knowable. On the contrary, only the naturalist can truly engage in creative thought on and beyond those boundaries, without her speculations declining into well-groomed wishful thinking.
22. Such a dogmatic approach perhaps, better corresponds to the term *scientism.* Van Fraassen is right when he writes that '[t]o the extent that scientific realism shades into scientism, it ... require[s] the sacrifice of the intellect, the

desperation of "*Credo ut intelligam*"' and that '[t]he alternative is to accept the challenge of intellectual maturity: to let your faith be not a dogma but a search, not an answer but a question and a quest, and to immerse yourself in a new world-picture without allowing yourself to be swallowed up' (in Hilgevoord 1994: 133). I share this desire not to forfeit our philosophical wits in the reassuring belief in a domesticated universe: however, the choice of intellectual maturity need *not* lead to a radical empiricist stance, but on the contrary can bolster a realist one.

23. Boucher and Vincent describe Collingwood's philosophy as 'an unusual variant of Idealist thinking, owing more to Hegel, Croce, Gentile, de Ruggiero and John Ruskin, than to English and Scottish Idealism' (2000: 185).
24. But see A.N. Whitehead, who wrote, a few decades before Collingwood, that '[w]hen you are criticising the philosophy of an epoch, do not chiefly direct your attention to those intellectual positions which its exponents feel it necessary explicitly to defend. There will be some fundamental assumptions which adherents of all the variant systems within the epoch unconsciously presuppose. Such assumptions appear so obvious that people do not know what they are assuming because no other way of putting things has ever occurred to them' (1948: 49).
25. This Strawsonian term is employed with reference to Collingwood by Giuseppina D'Oro (2002) who correctly argues for a 'Kantian heritage' in Collingwood's understanding of metaphysics, both thinkers being concerned with the structural-conceptual conditions of our knowledge of things. The neo-Kantian interpretative key can be used to see how Collingwood's position can be placed in an interesting dialectic with that of his contemporary Rudolf Carnap. The early Carnap deflates the content of metaphysical claims in similar terms than Collingwood, claiming that '[t]he (pseudo)statements of metaphysics do not serve for *the description of states of affairs*, neither existing ones (in that case they would be true statements) nor non-existing ones (in that case they would be at least false statements). They serve for the expression of the general attitude of a person towards life (*Lebenseinstellung, Lebensgefühl*)' (in Sarkar 1996: 78). The later Carnap, on the other hand, believed that to historicise the Kantian *a priori* structures amounts to recasting them as valid from within pragmatically adopted 'linguistic frameworks' (see Carnap 1950; for a reconsideration of Carnapian logical positivism as a thoroughly post-Kantian project see Friedman 1999 and for an analysis of logical positivism as a 'semantic tradition' of thinkers concerned with *a priori* structures but rejecting the constitutive powers of mind see Coffa 1991). However, neither unaccounted-for Collingwoodian historical change nor Carnap's pragmatic choices (without considerations of metaphysical adequacy) will satisfy the realist seeking a *rational assessment* and *evaluation* of presuppositions/paradigms/frameworks.
26. Note that I am not adopting either Collingwood or Bachelard as *realist* models. It is debatable whether either of them would recognise themselves in the definition of scientific realism I offered above (on Bachelard see Gutting 1987 for a realist interpretation and McArthur 2002 for an anti-realist one). I do, however, think it is the realist's duty to engage in the kind of historical considerations they were interested in.

27. Gutting notes that Bachelard's 'account of scientific change allows [him] to reject the *continuity* of science and still accept its *progress*' (1987: 62).
28. We could (echoing Marx, Engels and Mao) define such fervent process of knowledge-revision as a *permanent* (epistemological) *revolution*. Balibar (1978: 211) indeed writes that '*[t]he very category of knowledge, is the expression of this transition, of this antithetical relation, of this process of rupture*'.
29. 'We believe, indeed, that scientific progress always manifests a break [*rupture*], perpetual breaks, between common knowledge and scientific knowledge, as soon as one examines an evolved science – a science which, by virtue of these breaks, bears the mark of modernity' (Bachelard 1963: 207; my translation).
30. Althusser similarly exploited Bachelard's concept of epistemological breaks in order to index the emancipation of a scientific thought from a pre-scientific ideology arguing that

    > [t]here are false ideas about science, not simply in the heads of philosophers but in the heads of scientists themselves: false 'obviousnesses' that, far from being means of making progress, are in reality 'epistemological obstacles'... They are non-scientific, ideological ideas and representations. They form what we will provisionally call scientific ideology, or the ideology of scientists. A philosophy capable of discerning and criticizing them can have the effect of drawing the attention of scientists to the existence and efficacy of the epistemological obstacle that this spontaneous scientific ideology represents: the representation that scientists have of their own practice, and of their relationship to their own practice... Scientific ideology (or the ideology of scientists) is inseparable from scientific practice: it is the 'spontaneous' ideology of scientific practice. (1990: 88)

    For a larger-scale analysis of the post-Bachelardian tradition of Marxist epistemology see Lecourt (1975).
31. Fundamental references for the ways metaphors condition our usage and creation of concepts are Black (1962); Derrida (1982); Lakoff and Johnson (1980); Hesse (1988). An interesting comparison can be made here with the debate between Richard Boyd and Thomas Kuhn on the role of metaphor in the sciences in Ortony (1979). While Boyd defended the realist cause by arguing that 'the use of metaphor is one of many devices available to the scientific community to accomplish the task of accommodation of language to the causal structure of the world', i.e. 'the task of introducing terminology, and modifying usage of existing terminology, so that linguistic categories are available which describe the causally and explanatorily significant features of the world' (in Ortony 1979: 483), Kuhn took issue with the Platonic image of cutting 'the world at its joints', repeatedly employed by Boyd, arguing that no continuity of reference to natural kinds nor any theory-neutral language can be discovered in science. Clarifying that both he and Boyd qualify as 'unregenerate realists' (in Ortony 1979: 539), he nevertheless bemoans Boyd's *ontological* extension to the 'joints' of the 'world' metaphors, for which he argues – through pessimistic meta-induction – there is no historical warrant. 'Boyd's world with its joints seems to me', Kuhn concluded, 'like Kant's "things in themselves", in principle unknowable. The view toward which I grope would also be Kantian but without "things in themselves" and with categories of the mind which could change with time as the accommodation

of language and experience proceeded. A view of that sort need not, I think, make the world less real' (in Ortony 1979: 540). I cannot imagine how this almost Hegelian stance can deliver a 'real' world in any robust sense of the term.

32. As Norris explains, for Bachelard 'knowledge could never have progressed beyond the stage of naïve sense-certainty were it not for this capacity of critical thought to revise and modify its own preconceptions in response to new challenges or obstacles' (2000: 110). Importantly, this emphasis on conceptual revision is not to be read as the psychologistic expectation of singular, ineffable flashes of insight from the individual scientist: of the two orientations running through French philosophy that Michel Foucault famously described (in Canguilhem 1978: x) as 'a philosophy of experience, of sense, and of subject', on the one hand, and 'a philosophy of knowledge, of rationality and of concept', on the other, Bachelard (like Foucault himself, his indirect intellectual heir through Canguilhem – and indeed like Badiou) belongs squarely to the second. It is not, for Bachelard, a matter of explaining how empirical subjectivity operates, but to explicate rational patterns of discovery.
33. To quote a relevant passage from Sellars to which Bachelard would have given, I suspect, his unconditional agreement,

    > [t]he essence of scientific wisdom consists in being uncertain about what is certain, in a readiness to move from one conceptual frame to another. For not only can we be caused to modify our linguistic frame, we can deliberately modify it – teach ourselves new habits – *and give reasons for doing so*. Now, the use of a conceptual frame is the awareness of a system of logical and extra-logical necessities. The essence of scientific wisdom, therefore, lies in being tentative about what one takes to be extra-logically necessary. (1963: 319; emphasis added)

34. Note Sellars's conviction that

    > the human mind is not limited in its categories to what it has been able to refine out of the world view of primitive man, any more than the limits of what we can conceive are set by what we can imagine. The categories of theoretical physics are not essences distilled from the framework of perceptual experience, yet, if the human mind can conceive of new categories, it can also refine the old; and it is just as important not to overestimate the role of creativity in the development of the framework in terms of which you and I experience the world, as it is not to under-estimate its role in the scientific enterprise. (1963: 10)

35. The contested issue of progress is what sets apart the Bachelard-inspired approach I am taking here from the closely related project of 'archaeological' intellectual history of Michel Foucault. Like Collingwood and Bachelard, Foucault had a special interest in those revolutions of thought which modify the presuppositions we bring to bear on both empirical inquiry and social organisation. Foucault concentrated his analysis on 'the epistemological field, the *episteme* in which knowledge, envisaged apart from all criteria having reference to its rational value or to its objective forms, grounds its positivity and thereby manifests a history which is not that of its growing perfection, but rather that of its conditions of possibility' (2002: xxiii–xxiv). A first difference between the Bachelardian project and Foucault's own

'archaeological' phase is the broader temporal scope of Foucaultian *epistemes* (and consequently the relative rarity of conceptual discontinuities between them). In *The Order of Things* Foucault's analysis is organised around three large historical periods (the pre-Classical (broadly, the Renaissance), the Classical and the Modern) and the 'two great discontinuities' which separated them, the first 'roughly half-way through the seventeenth century' and the second 'at the beginning of the nineteenth century' (2002: xxiv). The second, much more important difference is the overt rejection, on Foucault's side, of any concept of rational progression (this putting Foucault more in the Collingwoodian camp). Regarding the shift in *epistemes* Foucault is quick to clarify his stance: '[n]ot that reason made any progress: it was simply that the mode of being of things, and of the order that divided them up before presenting them to the understanding, was profoundly altered' (2002: xxiv). I cannot do justice to the nuanced distinctions between these positions here, but let me note that – while it is certainly possible to find in Foucault's analysis of the productive power of discourse a form of linguistic idealism (much more explicitly than Bachelard's emphasis on *mutual* influence between the rational mind of the scientist and reality) – Foucault's scepticism towards Enlightenment conceptions of rational progress is more complex than the pre-packaged form in which lesser commentators deliver it to impressionable undergraduate intellects (see Allen 2003 for an excellent reappraisal of Foucault's Kantianism). Nowhere more strongly does Halperin's admonition apply, lamenting how 'the almost ritualistic invocation of [Foucault's] name by academic practitioners of cultural theory, has had the effect of reducing the operative range of his thought to a small set of received ideas, slogans, and bits of jargon that have now become so commonplace and so familiar as to make a more direct engagement with Foucault's texts entirely dispensable' (1998: 93–4). Limiting myself to gesturing towards the crucial issue here, Foucault's rejection of 'reason' is the *anti-humanist* rejection of self-aware, autonomous, free (in a word, *Sartrean*) rational *subjectivity* progressing 'in its empty sameness' through history. Hence Foucault's admonition that 'just as we must free ourselves from the intellectual blackmail of "being for or against the Enlightenment", we must escape from the historical and moral confusionism that mixes the theme of humanism with the question of the Enlightenment' (in Rabinow 1984: 45). Foucault's critical interest in 'power' as the set of practices of social control over the individual subject (from behavioural monitoring to bodily normalisation), while certainly motivated by a pessimistic outlook on the consequences of humanistic universalism, does not entail anti-progressive consequences across the board (especially not if read as gingerly crossing the divide between *social* and *natural* sciences), and certainly does not motivate scepticism towards the progress of science in understanding reality (had he survived his illness, I suspect that Foucault would have been rather interested in the kinds of progress made by neurology and cognitive science in the last two decades in unveiling the illusory nature of our self-present subjectivity). As I will observe below (and as I did in my reference to the Stanford School) the recognition of how our conceptual categories at times arbitrarily force order upon disordered realities need not be a prelude to an anti-realist/relativist conclusion. If there is one Enlightenment-inspired motto that I am convinced Foucault would have endorsed it is that we are *never done* improving our epistemic practices.

36. As Spinoza put it, 'all the prejudices which I intend to mention here turn on this one point, the widespread belief among men that all things in Nature are like themselves in acting with an end in view…Nature has no fixed goal and…all final causes are but figments of the human imagination' (2002: 239, 240; Appendix I).
37. Ray Brassier offers a very clear description of a responsible commitment to naturalism, rejecting its interpretation as 'a metaphysical doctrine engaging in an ontological hypostasis of entities and processes postulated by current science' but endorsing it as 'an epistemological constraint stipulating that accounts of conception, representation, and meaning refrain from invoking entities or processes which are in principle refractory to any possible explanation by current or future science'. He further suggests that the realist philosopher should grant 'maximal (but not, please note, incorrigible) authority to the scientific representation of the world while acknowledging that science changes its mind about what it says there is' (in Bryant et al. 2011: 49, 64).
38. A recent magisterial account of the pragmatics of concept revision in the natural sciences is Wilson (2006).
39. Adrian Johnston summarises this sentiment well when he observes that today 'one can step back and see the natural sciences themselves developing out of their own resources a sense of their limitations *vis-à-vis* the things that, philosophically speaking, we are interested in. We can begin to account for how the sciences, on their own terms, are necessarily incomplete and that they can actually pinpoint the ways in which they're incomplete' (2011: 167). This stance has a peculiarly controversial 'continental' history which, without suggesting a direct line of descent, should be mentioned precisely to underline its discontinuities. Martin Heidegger's attitude towards the sciences is mostly remembered through his notorious dictum that '[s]cience does not think' (1968: 8). However, it is instructive to contextualise this claim by referring to his treatment of the subject in what follows. In *What is Called Thinking*, Heidegger writes that

    > [b]y way of history, a [human] will never find out what history is no more than a mathematician can show by way of mathematics – by means of his science, that is, and ultimately by mathematical formulae – what mathematics is. *The essence of their sphere* – history, art, poetry, language, nature, man, God – *remains inaccessible to the sciences*. At the same time, however, the sciences would constantly fall into the void if they did not operate within these spheres. The essence of the spheres I have named is the concern of thinking. *As the sciences qua sciences have no access to this concern, it must be said that they are not thinking*. (1968: 32; emphases added)

    However, Heidegger then proceeds to qualify his claim, explaining that

    > [o]nce this is put in words, it tends to sound at first as though thinking fancied itself superior to the sciences. *Such arrogance, if and where it exists, would be unjustified*; thinking always knows essentially less than the sciences precisely because it operates where it could think the essence of history, art, nature, language – and yet is still not capable of it. *The sciences are fully entitled to their name, which means fields of knowledge, because they*

have infinitely more knowledge than thinking does. And yet there is another side in every science which that science as such can never reach: the essential nature and origin of its sphere, the essence and essential origin of the manner of knowing which it cultivates, and other things besides. The sciences remain of necessity on the one side. *In this sense* they are one-sided, but in such a way that the other side nonetheless always appears as well. (1968: 32–3; emphases added)

It is not my aim to offer a detailed apology for Heidegger's characteristically cryptic statements, but it is useful to point out how it would be mistaken to assume that this complex stance *vis-à-vis* the sciences – far less disdainful than it is often, anecdotally, presented to be – left no trace in subsequent decades of 'continental' philosophy of science, thus making it harder to dismiss this tradition as ineluctably compromised with an anti-realist, if not anti-scientific, posture.

40. Cf. Peirce *CP*: 1.129.
41. Seibt (2000) convincingly presents Sellars as engaged in what she calls a 'projective metaphysics', a science-informed 'projective category construction', revising inherited categorical frameworks.
42. This is no concession to anti-realism. The import of Sellars's neo-Kantian denunciation of the foundationalist 'myth of the Given' (in its most poignant form, the rejection of '*the idea that the categorial structure of the world – if it has a categorial structure – imposes itself on the mind as a seal imposes an image on melted wax*' (1981: 12)), which underwrites these claims, is precisely that of engendering a (robust yet fallibilist) *realism* capable of responsibly acknowledging and examining the structure of conceptual inferences that make genuine, cognitively relevant, perception – rather than mere *sensation* – possible. Brandom sums it up by explaining that 'observational concepts...can be thought of as *inference laden* [but] [i]t does not follow...that they are for Sellars for that reason also *theory laden*' (2002: 353). Brassier articulates the necessity of these sorts of epistemological considerations thus: '[t]hought is not guaranteed access to being; being is not inherently thinkable. There is no cognitive ingress to the real except through the concept. Yet the real itself is not to be confused with the concepts through which we know it. *The fundamental problem of philosophy is to understand how to reconcile these two claims*' (in Bryant et al. 2011: 47; emphasis added). This approach, I would add, amounts to a realism (neither *internal* nor *external*, to use Putnam's all-too-rough division) sympathetic to the socio-historical study of extra-scientific influences on the natural sciences that goes under the various names of Sociology of Scientific Knowledge (SSK), Science and Technology Studies (SST) or History and Philosophy of Science (HPS). The realist should welcome the critical examination of science (in its theoretical and practical aspects) seeking social biases of various kinds (political, class, gender or race), either in its present or past form, as precious allies in the process of purifying and sharpening our concepts from delusory biases and vestigial beliefs.
43. Thus, while I agree with Nagel's claim that materialist reductionism and the Darwinian evolutionary picture are 'assumption[s] governing the scientific project rather than...well-confirmed scientific hypothes[es] (2012: 11), yet I would add that they are not *unsupported* assumptions. Indeed they are

the *best* assumptions we as a species have collectively developed over the course of centuries of natural sciences in order to make coherent sense of the universe.

44. The term was coined by Nancy Nersessian to describe an analysis of case studies in the history of science that puts special emphasis on detecting the shifts in conceptual frameworks which are at work therein, and that applies to them models for the acquisition and modification of knowledge drawn from cognitive science. See Nersessian in Giere (1992) and Nersessian (2008).
45. And note that only an overtly optimistic rational reconstruction of the history of science would describe these criteria as having been just *rules*, since it often is more accurate to talk about *values*.
46. As Brassier puts it:

    > [m]eaning cannot be invoked either as originary constituent of reality (as it is for Aristotelian essentialism) or as originary condition of access to the world (as it is for Heidegger's hermeneutic ontology): it must be recognized to be a conditioned phenomenon generated through meaningless yet tractable mechanisms operative at the sub-personal (neurocomputational) as well as supra-personal (sociocultural) level. This is a naturalistic imperative. (in Bryant et al. 2011: 48–9)

47. Nietzsche clearly discerned this danger when he noted that to reverse anthropocentric values of order and meaning is *still* an unwarranted ascription laden with value judgements, 'an anthropomorphism bearing a reproach. But how could we reproach or praise the universe!' (2001: 109).

4 Structural Realisms

1. See Carnap in Sarkar (1996) for a classic programmatic statement.
2. Carnap argued that

    > [a]ny physical theory, and likewise the whole of physics, can in this way be presented in the form of an interpreted system, consisting of a specific calculus (axiom system) and a system of semantical rules for its interpretation; the axiom system is, tacitly or explicitly, based upon a logico-mathematical calculus with customary interpretation. It is, of course, logically possible to apply the same method to any other branch of science as well. (1939: 60)

3. I am here omitting the internal debate between the positivists themselves (along the Schlick–Neurath axis) regarding the proper interpretation of protocol statements' intention (a phenomenalist one, expressed in sense data language and a physicalist one expressed in thing-language). See Uebel (1992) for an in depth-analysis.
4. Considering this positivists' desired removal of reference to unperceivable entities, it is somewhat ironic that Frege – the thinker who singlehandedly delivered a new formal apparatus uniquely capable of grounding the kind of demystifying logical analysis of language the logical positivists were so keen to pursue – compared his *Begriffsschrift* (his formal notation for the expression of concepts) to a microscope, necessary when 'scientific purposes place great demands on sharpness of resolution' and thus 'the eye turns out to be inadequate' (Frege in Beaney 1997: 49).

5. Of course, the scientific realists' counterattack was not the only cause of the demise of logical empiricism. Both internal unresolved tensions (the unverifiable status of the verifiability criterion) and critiques from other quarters jointly contributed to its demise: chiefly, Popper's questioning of the strict line between science and metaphysics drawn by Carnap (see Popper 1962: Chapter 11), Kuhn's insistence on the positivists' idealised and unhistorical understanding of scientific practice (see Kuhn 2012 [1962]), Hanson's Wittgensteinian reflections on the process of scientific theory-change and attack on the theory–observation distinction (1958), and of course Quine's infamous rejection of the two dogmas of empiricism (the analytic–synthetic distinction and the verificationist theory of meaning; see Quine 1953: Chapter 2). For a critical questioning of the common narrative of how Kuhn singlehandedly undermined the positivist program see Reisch (1991), and for a nuanced analysis of the socio-political pressures at play in Cold War, McCarthyist America which led to the extinction of the logical empiricist movement see Reisch (2005). For an assessment of Hanson's under-examined influence on anti-realism in philosophy of science see Norris (2004). Studies on Quine's relation to logical positivism and to Carnap in particular are too numerous to be listed: a concise general entry point is Isaacson in Gibson (2004), while an excellent recent re-examination of the crucial Quine–Carnap debate is Price in Chalmers et al. (2009). It should be added that the demise of logical positivism did not merely foster the development of scientific realism but, on the broader philosophical scene, allowed for the birth – largely through the pioneering work of Quine and Strawson, rehabilitating talk of 'ontological commitments' and 'descriptive metaphysics' – of what are now known as analytic ontology and metaphysics (see Quine 1953 and Strawson 2003).
6. See Salmon 1984.
7. See also Boyd arguing along similar lines that for the realist 'experimental evidence for a theory which describes causal relations between "theoretical" (that is, unobservable) entities is evidence not only for the correctness of the observational consequences of the theory, but is also evidence that the particular causal relations in question explain the predicted regularities in the behavior of observable phenomena' (1973: 1). A relevant debate regarding scientific realism took place between Sellars and van Fraassen in the mid-1970s (Sellars 1976; van Fraassen 1975, 1976). Its interest lies especially in its representation of a generational shift, where a young van Fraassen objects to Sellars's realism by relaxing the verificationist empiricism of the logical positivists, thus beginning to lay out the tenets of the constructive empiricism he will fully deploy from the 1980s onwards. One can discern in van Fraassen's refusal to treat Sellars's 'manifest image' as guided by the same theoretical structure and explanatory demands as the scientific one, the deepest motivations for his anti-realist stance.
8. Smart also voices agreement with Feyerabend and argues that a phenomenalist or positivist stance would actively oppose scientific *progress* by preferring observationally more accurate theories against those newer theories with a broader explanatory ambition but with (initially) minor observational accuracy. For Smart, then '[i]t is easy to see that positivism is easily allied with dogmatism... [A] physicist who adopts an instrumentalist position will

be satisfied with the prevailing point of view, and will strongly resist any attempt to produce alternatives' (1963: 40). Leplin (1997) follows this line by focusing on how realism is best defended as the only position allowing for predictive novelty.

9. Josephson and Josephson claim that '[c]onsidering its apparent ubiquity, it is remarkable how overlooked and underanalyzed abduction is by almost 2,400 years of logic and philosophy' (1994: 27). Indeed it is only in relatively recent times that abductive reasoning was put at the centre of philosophical focus. The classic exposition of IBE is due to Harman (1965), in an article responsible for the diffusion of the term 'inference to the best explanation'. The most thorough critical defence of IBE is Lipton's, trying to disentangle and do justice to its complex structure (including the often unclear concept of 'explanation' itself), in the belief that 'it still remains more of a slogan than an articulated account of induction' (2004: 57). Interdisciplinary approaches to abduction and IBE, related to scientific reasoning, computer science and clinical diagnosis, are Josephson and Josephson (1994), Aliseda (2006) and Magnani (2001), the latter offering a particularly interesting interpretation of abductive reasoning as a generalised rational model of creative discovery, differentiating between '1) abduction that only generates "plausible" hypotheses (*selective* or *creative*) and 2) abduction considered as *inference to the best explanation*, which also evaluates hypotheses' (2001: 19).
10. Induction has an ampliative nature to the extent that any inductive generalisation should allow reference to the properties of hitherto unobserved cases, but offers nothing in the way of an *explanation* (hence Hume's problem). According to standard logic, then, abduction is an unsound inference rule, from which derives its *non-monotonic* character. As Magnani observes '[i]t is important to allow the guessing of explanations for a situation, in order to discount and abandon old hypotheses, so as to enable the tentative adoption of new ones, when new information about the situation makes them no longer the best' (2001: 24).
11. Note here the interesting tension between McMullin's choice of words like 'risk', 'venture' and 'strict limits' with van Fraassen's characterisation of the realist as a feeble-minded individual incapable of accepting the contingency of the real and vainly seeking metaphysical reassurances. This conflict is central to my arguments: the empiricist position is not the only coherent choice to avoid intellectual laziness. Indeed there can be realist programmes pursuing metaphysical knowledge without the safety net of a stable mooring of reality.
12. See Psillos (1999: Chapter 4) for a popular IBE-based argument for scientific realism.
13. For naturalist critiques of the cogency of IBE for science see Maddy (2007), arguing that the naturalist need not 'step back' into a second-level epistemic stance (over and above the immersion in scientific practice), and Ladyman and Ross (in Ladyman et al. 2013), arguing that the 'best explanation' targeted by IBE is all too often read as a 'most intuitively satisfying explanation', hence putting a premium on those common-sense metaphysical intuitions which the naturalists vigorously rejects. Chakravartty (in Ladyman et al. 2013) offers an excellent rejoinder to these arguments. See also the debate on IBE between Psillos (1996, 1997) and Ladyman et al. (1997).

14. Some commentators have argued that William Whewell's inquiries into scientific epistemology and in particular his innovative notion of 'consilience of inductions' can be considered as an antecedent treatment of IBE. Enrnan McMullin, for example, wrote that 'when a "consilience of inductions" occurs, when hitherto unrelated areas of inquiry fall together under a single hypothesis, this can ... convince us of the truth of the hypothesis. Consilience involves, then, both enlargement of scope and simplification of structure' (1992: 85). Laura Snyder, however, has convincingly shown the two inferential processes to be heterogeneous since

    [w]hat is important in the case of consilience ... is not merely that a single causal mechanism or law can explain or account for different event kinds (as the inference to the best explanation allows), but rather that separate lines of induction lead from each event kind to the same causal mechanism or law, that there is a convergence of distinct lines of argument. (2006: 175)

    This is more than a marginally interesting historical note considering my interest in flagging theologically skewed inference patterns: Whewell's understanding of consilience, as correctly reconstructed by Snyder, is directly supported by his theological commitments; the convergence of different instances of 'discoverer's induction' (in his words) is ultimately explained by a divinely established natural order. Consider how Whewell writes (after having discussed Kepler's struggle to 'discern' the laws of elliptical planetary motion), that

    [w]e may readily conceive the satisfaction and delight with which, after this perplexity and struggle, the discoverer finds himself in light and tranquillity; able to look at the province of nature which has been the subject of his study, *and to read there an intelligible connexion, a sufficing reason,* which no one before him had understood or apprehended. This step so much resembles the mode in which one intelligent being understands and apprehends the conceptions of another, that we cannot be surprised if those persons in whose minds such a process has taken place, have been most ready to acknowledge *the existence and operation of a superintending intelligence, whose ordinances it was their employment to study*. When they had just read a sentence of the table of the laws of the universe, they could not doubt whether it had had a legislator. When they had decyphered there a comprehensive and substantial truth, they could not believe that the letters had been thrown together by chance. They could not but readily acknowledge that *what their faculties had enabled them to read, must have been written by some higher and profounder mind*. (1862: 264; emphases added)

    Unlike Whewell's discoverer's induction, the epistemic efficacy of which ultimately depends on the creative act of God, abductive processes put a premium on the inventiveness of the rational agent, a creative inference pattern that allows us to bootstrap our own reason in the attempt to grasp reality.
15. Prediction is a mode of inductive generalisation concerned with projecting into the future the *same result* of present observations, not with the creation of a *new* hypothesis. As Josephson and Josephson explain: 'predictions and abductions are two distinct kinds of plausible inference. Abductions go from

data to explanatory hypothesis; predictions go from hypothesis to expected data' (1994: 25).

16. For an interesting discussion on the validity of this argument see the debate between Leplin (defending it) and Kukla and Walmsley (criticising it) in Hitchcock (2004).
17. There are those who do not consider IBE to be an *argument* at all. Van Fraassen demotes its status to that of an 'intuition', blind to the actual history of science and grounded either upon the claims of philosophically naive scientists or 'putatively found in the commonsensical [human] in the street' (2006: 296). According to van Fraassen 'there is an accumulation of knowledge through science, but it is knowledge about the observable phenomena' and therefore 'the success of science is *not a miracle*, because in any theoretical change both the past empirical success retained and the new empirical success *were needed as credentials* for acceptance' (2006: 298–9). The empiricist stance, then, reverses the explanatory order, arguing that it is empirical success that must be taken as a brute fact upon which theoretical change can be justified, and that no metaphysical justification must be sought for empirical success. As I noted in the previous chapter, when the realist–anti-realist debate is cast as a clash between such basic commitments I do not think that any intra-philosophical answer can be given to settle the issue once and for all. Only meta-philosophical commitments can tilt the balance towards either position. This, incidentally, is the reason why van Fraassen's position, committed like the realist to naturalist respect for the inner workings of science, is such a mighty opponent for the realist.
18. See Duhem (1991 [1914]).
19. Before I proceed to examine structuralist retorts to this sceptical argument, I would endorse Ian Hacking's *historical* riposte. Hacking notes that '[c]ontrary to the themes of Karl Popper and Thomas Kuhn [and one could add Laudan], namely refutation and revolution, a great deal of modern science is stable' and argues that '[f]uture historians of the history and philosophy of the sciences may suggest that Popper and Kuhn worked in unusual times' (1999: 33, 85), i.e. the uncharacteristically revolutionary first half of the twentieth century. In other words, Hacking does not question the validity of the historical case studies buttressing sceptical arguments but the validity of the induction from them to the future of science. Of course, reliance on the present stability of science should be endorsed 'with caution and humility' (1999: 85), but it would be a mistake to reject the possibility that a certain subset of contemporary scientific theories are going to survive throughout the evolution of science. Human beings have played the epistemic game of scientific inquiry for a relatively short time (as compared with the age of the universe they endeavour to understand) and to base predictions of science's future on a handful of centuries of cognitive struggle would be a hasty generalisation.
20. Worrall's approach is not the *only* way to respond to PMI. See for example Psillos's (1999: Chapter 5) *'divide et impera'* strategy.
21. Worrall (1989). See also Worrall (1994) and Worrall in Psillos and Curd (2008).
22. The notion of structure can be mathematically described as composed by a non-empty set U of objects or relata (the domain of the structure) and a non-empty set R of extensional relations on U (including one-place relation, i.e.

monadic/intrinsic properties). Such a structure is abstract in the sense that 'objects' are nothing but featureless place-holders or nodes of relations. The same *structure* can be concretely instantiated in different concrete *systems,* where the same relations are preserved between different individuals occupying the abstract nodes. I will return to this type–token distinction below when addressing the issue of mathematical structuralism.

23. Structural realism has been either the main topic or an object of discussion in over 50 publications (including journal articles, monographs and PhD theses) in the 15 years between 1998 and 2013. It is undoubtedly a burgeoning field.
24. This loose definition of structuralism will allow me, in Chapter 5, to argue that the *aporiae* or ambiguities that other forms of structuralism have been demonstrated to incur (mostly from twentieth-century philosophy of language and mathematics) can also be expected to emerge in the structuralism defended in the philosophy of science, of course with much deeper metaphysical repercussions. For a thorough diagnosis of the self-undermining nature of the structuralist project in analytic philosophy of language see Livingston (2008). For a classic statement of the *aporiae* of Saussurrean structuralism see Derrida (1982).
25. For contemporary examples of anti-realist structuralism see van Fraassen (2006) and Bueno (1999, 2000). Cassirer, Schlick, Eddington and Carnap have all held structuralist positions different from the current realist one: Gower (2000) offers an insightful historical analysis of two different avenues to structuralism – one of neo-Kantian and the other of logical empiricist origin – in the thought of the first two, French (2003) examines Eddington's structuralist philosophy of physics, while Psillos (1999) and Friedman (2009) offer a structuralist reading of Carnap's Ramsey-sentence approach to theoretical objects. A thorough reconstruction of the history of structuralism as a whole in the philosophy of science is in Votsis 2004 (Ch. 2). A comprehensive and fine-grained review of the present landscape of possible arguments for and against structural realism is Frigg and Votsis (2011). My own exposition here will not aim at the same level of detail that Frigg and Votsis offer. I shall concentrate on more specific elements in (and consequences of) structural realism, and I will cite relevant sources as my argument progresses. I should also note that in taking Worrall as my starting point I am choosing to reconstruct structural realism according to what Psillos (2009) described as its 'downward path' (a naturalist structuralism, reached by starting from well-formed scientific theories and proceeding 'down' towards the epistemological and ontological interpretations of their results) and I will not examine in detail the 'upward path' (an empiricist structuralist stance derived from epistemological reflections on our perceptual capabilities) which Psillos identifies as originating in the work of Russell and taken up by Maxwell (see in particular Russell 1992, 2009; Maxwell in Radner and Winokur 1970). For a defence of the 'upward path's' ability to deliver inferential knowledge of the external world, against Psillos's objections, see Votsis (2005).
26. Recently, an extension of the structuralist approach towards the sciences other than physics has been attempted, including biology (French 2011, 2014), economics (Ross 2008) and the social sciences as a whole (Kincaid 2008; Ladyman and Ross 2007).

27. Poincaré is usually viewed as a paradigmatic case of a conventionalist due to his positions vis-à-vis mathematics and geometry. However, Elie Zahar (2001) has convincingly shown that his structuralist approach in the philosophy of science is best understood as a realism, his contention being that 'in a unified and highly confirmed theory, the basic relations mirror some objective reality while the relata remain inaccessible to human knowledge' (2001: 38).
28. See also Weyl's markedly Kantian take on the same issue, stating that

    > [a] science can only determine its domain of investigation up to an isomorphic mapping. In particular it remains quite indifferent as to the 'essence' of its objects... It would be folly to expect cognition to reveal to intuition some secret essence of things hidden behind what is manifestly given by intuition. The idea of isomorphism demarcates the self-evident insurmountable boundary of cognition. (1949: 25, 26)
29. Worrall thus does not require an exact preservation of structure for his argument but only that the old structure be embeddable as a limit case into the new one. For a classic definition of the correspondence principle see Post (1971).
30. I borrow the term from Rae Langton's important study (1998), where she defends a scientific realism-compatible interpretation of Kant – where knowledge of things in themselves is abandoned in favour of that of 'mere relations' – in broad agreement with Worrall–Poincaré structural realism.
31. Thus Worrall (1994) rejects the more robustly realist idea that, in the case study of the transition between Fresnel's ether and Maxwell electromagnetic field, the two scientists somehow referred to the same *underlying entity* all along. This 'in-between' status of Worrall's stance is most clearly presented when he claims that '[Fresnel's] theory was more than empirically adequate, but less than true; instead it was *structurally correct*' (1994: 340).
32. Floridi (2008: 223) comments that 'as a form of syntactic realism, eliminativist OSR betrays the original Kantian polarisation between knowable phenomena and unknowable noumena, which lies at the roots of SR. It starts resembling a metaphysically more problematic form of absolute idealism, according to which "whatever is real is structural and whatever is structural is real", to paraphrase Hegel.' I disagree with Floridi's conclusion, but the rationalist-Hegelian challenge is indeed one which must be met. See Chapter 5.
33. See in particular Bueno et al. (2002); French (2000, 2006, 2014); French in Castellani (1998); French and Ladyman (2003a, 2003b); French and Ladyman in Bokulich and Bokulich (2011); Ladyman (1998, 2001, 2005, 2008, 2009); Ladyman and Ross (2007); Ladyman and Ross in Ladyman et al. (2013).
34. Chakravartty dryly comments that '[t]he kind of knowledge [ESR] claims regarding unobservables falls short of anything resembling scientific realism' (2007: 39). This is perhaps too strong a claim, one which should be moderated by a reading of ESR in a Kantian light. Ladyman and French (in Bokulich and Bokulich 2011: 27) argue that two versions of ESR could be disentangled: '[a] ccording to ESR1 there are [unobservable individual] objects but we cannot know them, and according to ESR2 there may or may not be such objects, but we cannot know either way, and if there are such objects we cannot know them.' Morganti (2004) defends an 'agnostic' form of ESR (one which fits Ladyman and French's ESR2), arguing that it is 'better conceived as a position according

to which there could be something more beyond structures [then it could be read] as one which asserts that there surely is something, but we cannot know it' (2004: 101). As I noted, the debate of whether ESR can be interpreted as a realism largely overlaps with the vexed question of the possibility of a realist reading of Kant's transcendental idealism, unsurprisingly when we consider the historical lineage of this position (see Poincaré's passages quoted above). So Langton (1998) seems to describe an ESR-compatible, 'realist' Kant when she argues that Kantian noumena are to be interpreted as entities considered in their intrinsic, non-relational properties, unknowable to the extent that our time- and space-bound cognitive capacities can only deliver us knowledge of their secondary, relational properties. On the other hand, van Fraassen (2008) couches his discussion of contemporary structuralism, and his own empiricist variety of it, within the Kantian-inspired tradition of the *Bildtheorie* of science (an approach deriving from the discussions between Planck, Hertz and Boltzmann and continued by Duhem and Poincaré) describing scientific theories as *representations* of parts of reality, that is *non-mimetic* representations, van Fraassen argues, just instrumentally useful taxonomies granting no access to things in themselves. Once again, it seems to me fruitful to see the dialectic between ESR and OSR as structurally (!) similar to that between transcendental idealism and post-Kantian reactions to it.

35. This form of ontological parsimony (or wielding of Occam's razor) is purely *qualitative* (see Lewis (1973: 87) on qualitative and quantitative parsimony). This will be important to keep in mind when I come to Ladyman and Ross's notion of a plurality of 'patterns'.
36. The PII, commonly traced back to Leibniz, who makes large use of it in his metaphysics, is variously formulated but amounts to the thesis that if two distinct entities A and B share all of their properties, then A and B are identical. The negative formulation is somewhat clearer, stating that there cannot be two entities sharing the very same properties. For a classic critical analysis of the PII (and the infamous counter-example to it involving identical iron spheres) see Black (1952). For a contemporary, neo-rationalist defence of this principle (against Black's *reductio*) see Della Rocca (2005).
37. See French in Castellani (1998). Roughly put, quantum mechanical systems possess a form of symmetry called *permutation invariance*: if the particles' 'labels' are exchanged, the states of the quantum system containing them remain unchanged. The formalism of the theory, French explains, 'can be taken to support two very different metaphysical packages, one in which the particles are regarded as "nonindividuals" in some sense and another in which they are regarded as (philosophically) classical individuals for which certain sets of states are rendered inaccessible' (in Castellani 1998: 95). See French and Redhead (1988) for an early discussion of this problem, concluding that, given permutation invariance, it is still possible to describe quantum particles as individuals violating (a certain form of) PII. For a critique of French and Redhead's conclusion see Pooley in Rickles et al. (2006). See and French and Krause (2006) for a book-length general examination of the concept of identity in twentieth-century physics. Saunders (2006) has questioned the effective violation of PII by quantum particles (fermions) by appealing to an alternative notion of 'weak discernibility'. See Ladyman and Bigaj (2010) for a riposte to Saunders.

38. The ontological interpretation of structures sets apart the project of OSR from van Fraassen's constructive empiricism. The kinds of underdetermination considerations from quantum mechanics motivating OSR are the same which van Fraassen took as justifying his infamous 'goodbye to metaphysics' (1991: 480). Ladyman and Ross argue that the metaphysical commitment that OSR must hold in order to differentiate itself from van Fraassen's empiricism is that towards 'mind-independent modal relations between phenomena' (2007: 128): structural relations not reducible to empirical regularities and not supervening on individuals but ontologically fundamental (see also French and Ladyman (in Bokulich and Bokulich 2011: 31)). Ladyman (2000) offers a critique of van Fraassen's empiricism (and in fact, any radical empiricism) precisely on the grounds of its inability to adopt any stance on issues of modality. We have seen in the previous chapter how van Fraassen takes the rejection of metaphysical necessity as a badge of honour for the empiricist. The line between OSR and constructive empiricism is therefore a thin one: both recognise the explanatory failure of contemporary metaphysics in the face of non-domesticable scientific results. While the OSRist goes on adopting an optimistic if heavily revisionary attitude towards metaphysics, the constructive empiricist rejects the need for explanations and prefers a cautious agnosticism.
39. The implicit assumption (enjoying contemporary scientific consensus) being that QFT represents our best scientific theory regarding the fundamental level of reality, restricting our ontological commitments to matter fields and gauge force fields (see Teller 1995 and Kuhlmann 2010 for philosophical introductions and interpretations). French and Ladyman argue that OSR is the only realist stance capable of offering a metaphysical answer to the question 'what is a field?': 'the field is the structure, the whole structure and nothing but the structure. The thought here is that we cannot describe the nature of the field without recourse to the mathematical structure of field theory' (2003a: 48). See also French's discussion of Wigner's and Weyl's work on group theory in quantum physics (in Butterfield and Pagonis 1999), underlying the idea that elementary particles can be identified with sets of quantities invariant under symmetry groups. Other parts of modern physics are often marshalled in favour of (or at least given interpretations compatible with) OSR, such as space-time theories and quantum gravity (see Dorato 2000; Esfeld and Lam 2008; Ladyman 2001; Ladyman and Ross 2007; Stachel in Rickles et al. 2006).
40. Since, as French and Krause (2006) argued, the notion of non-individual object can be made intelligible in the domain of quantum physics. See also French (2014: Chapter 2).
41. Notice, then, that if Worrall's ESR was motivated by his allegiance to historicist attention to the development of science, OSR (particularly in Ladyman and Ross's work) is *primarily* buttressed by their staunch physics-first naturalism. Esfeld and Lam (2008: 30) argue that three main arguments can be employed to justify the position of OSR: (1) an argument from coherence, since Worrall's distinction between knowable structures and unknowable intrinsic properties introduces an undesirable gap between metaphysics and epistemology; (2) an argument from parsimony: the wielding of Occam's razor I mentioned above; and (3) an empirical argument from quantum

mechanics: the argument from the underdetermination between individuals and non-individuals in QM and between particles and field in QFT. See Ainsworth (2010) for a critique of the cogency of the argument from underdetermination. Frigg and Votsis (2011: 265) raise an interesting objection to argument 1), in the form of a simple question: 'why should there be no gap between epistemology and ontology? Why should there be only things we can know about?' This is an important question, one with Kantian echoes, casting an ominous idealist shadow on OSR. It indeed deserves an answer capable of avoiding both transcendental arguments and theological metaphysics. I will return to it in Chapter 5.

42. Ladyman takes OSR to be a naturalist stance which deems the projection of common-sense entities into the unobservable sphere to be unwarranted. Science imposes the relinquishment of all conceptual chauvinism, since '[t]he demand for a individuals-based ontology may be criticised on the grounds that it is the demand that the structure of the mind-independent world be imaginable in terms of the categories of the world of experience' and indeed Ladyman understands his scientism as politically progressive to the extent that '[t]hose who are most enamoured by science are often also those who are most critical of prejudices of class, race and gender' (2009: 147). Zahar makes a similar point against anthropocentric reasoning writing that '[t]hough presented as a realist desideratum, picturability is in effect an anthropomorphic and even an idealist requirement; for it seeks to assimilate all physical objects to certain intentional entities and hence to mere precepts' (2001: 45). A thoroughgoing (science-informed) restructuring of our conceptual apparatus consisting in the replacement of the notion of individual substance with that of relational structure would amount to nothing less than a revision of a Collingwoodian absolute presupposition, or the abandonment of a Bachelardian epistemological obstacle, and this is (as I explained in Chapter 3) precisely the task of metaphysical thought.

43. In her survey of possible ontologies for QFT, Johanna Seibt observes that the entire Western ontological research programme remains to this day captive to the Aristotelian paradigm of substance ontology, and that '"Characteristic Aristotelian Presuppositions" (CAPs) – have acted as constraints on ontological theory formation throughout the history of the discipline' (in Kuhlmann et al. 2002: 62). She argues that these CAPs can be seen as defining what she calls (echoing Sellars's formulation) the *'myth of substance'* (in Kuhlmann et al. 2002: 64) – where the notion of 'substance' can be roughly subsumed under the four functional roles of 'independence, subjecthood, persistence, and ultimate determinateness' (in Kuhlmann et al. 2002: 65). Cf. Bachelard's discussion of the 'substantialist obstacle' including the *'myth of interiority'*, seen as 'one of those fundamental processes of unconscious thought that are the hardest to exorcise' (2002: 107). One of the tasks of metaphysics is, as I observed above, precisely that of *historically* identifying and *naturalistically* eradicating such vestigial beliefs. Historicism and naturalism joining forces as bottom-up approaches: creative metaphysical generalisations need to come from the depths of our best archives of empirical data of both history and science. For a recent collection of views on how the canonical notion of 'body', 'object' and 'individual' are irremediably overturned by the physical sciences see Castellani (1998).

44. These terms are derived from Psillos's own (more complex) taxonomy of OSR (Psillos 2009: 137). Floridi (2008: 222) adopts the terms eliminative structural realism (EOSR) and non-eliminativist ontic structural realism (NOSR) for, respectively, the rejection of individuals *tout court* and the rejection of individuals as non-decomposible. For a further classification of OSR in three variants and their subvariants, according to which among relations, objects and properties are considered ontologically primitive see Ainsworth (2010).
45. Chakravartty (2007: 76) offers a list of three possible guidelines to observe when proposing revisions of ontological categories (need, explanatory role and fewer primitives) and argues that ORS satisfies none of them. Chakravartty's principles, however, are arguably guided by too conservative an outlook on metaphysical speculation. In line with my considerations on the creative element of the naturalist stance, I think that metaphysical revision can be more daring than Chakravartty allows for. As Sellars put it:

    [t]he categories of theoretical physics are not essences distilled from the framework of perceptual experience, yet, if the human mind can conceive of *new* categories, it can also refine the old; and it is just as important not to over-estimate the role of creativity in the development of the framework in terms of which you and I experience the world, as it is not to underestimate its role in the scientific enterprise. (1963: 10)

46. Chakravartty defends his semi-realism as a sophisticated realist approach capable of uniting the best of standard scientific realism, limiting one's ontological commitment to what he calls *detection properties* (causal properties linked to the regular behaviour of our detection instruments) with the structuralist insights of ESR.
47. See French and Ladyman (2003a, 2003b); Cao (2003a, 2003b).
48. Esfeld and Lam go on to define their own revisionary version of OSR called *moderate structural realism*, proposing that 'there are objects, but instead of being characterized by intrinsic properties, all there is to the basic physical objects are the relations in which they stand... neither objects nor relations (structure) have an ontological priority with respect to the physical world' (2008: 31). Esfeld and Lam (in Bokulich and Bokulich 2011) subsequently offered an amended version of their moderate structural realism, pivoting on the new, Spinoza-inspired thesis that the distinction between objects and relations is to be regarded as *conceptual* and not ontological, thus rendering it meaningless to talk of ontological priority: 'properties, including relations, are modes, that is, the ways in which objects exist' (in Bokulich and Bokulich 2011: 157).
49. Perhaps the most interesting attempt in this latter direction, presented as an intermediate shade between ESR and OSR (in line with Chakravartty's and Esfeld and Lam's proposals), is offered by Cao in his 2003a, 2003b articles and his 2010 monograph under the label of *constructive structural realism* (CSR). Cao criticises Ladyman and French's OSR, arguing that 'what differentiates [CSR] from other versions of structural realism is that the reality of unobservable entity can be inferred from the reality of structure. Methodologically, this suggests a structural approach to unobservable entity' (2010: 6). As such, CSR is presented as more committed to the knowability (contra ESR's agnosticism) and existence (contra OSR's eliminativism) of non-structural, individual entities. To the extent that there is more to

reality than structures, Cao's position, said to take 'structural knowledge as an epistemic access to non-observable entities' (2003a: 16), becomes at times very close to a *standard* scientific realism, defending the heuristic role of structural descriptions while preserving a firm belief in the ontological priority of relata to their relations. Cao accepts this proximity, but argues that the crucial difference is that 'instead of conceiving the physical world as consisting of fixed natural kinds, [CSR] conceive[s] the identity (or nature) of physical entities in a structuralist way, and thus a flexible rather than fixed way. This flexibility has opened a vast conceptual space for accommodating radical ontological changes while maintaining a realist sense of continuity' (2003b: 68–9). So what motivates the 'constructive' part of his structuralist approach is the history-based argument that 'our knowledge of non-observable underlying entities in science is mainly constructed and frequently reconstructed through our structural knowledge of these entities' and, more generally, 'the attainment of objective knowledge at the level of underlying ontology can only be realized through a historical process of negotiation among empirical investigators, theoretical reasoners and metaphysical interpreters' (2003a: 16).

50. The problem of causation is most thoroughly dealt with by Ladyman and Ross (2007: Chapter 5) and French (2014: Chapter 8). Most generally, they doubt that the traditional metaphysical accounts of causation can or should be applied to discussions about science: a clear concept of cause should be drawn *from* science, not imposed from a philosophical outside. For such a naturalised approach to causality see Ismael in Ladyman et al. (2013) and Norton in Price and Corry (2007).
51. The Principle of Naturalistic Closure (PNC) is one of Ladyman and Ross's (2007) guiding commitments. It states that '[a]ny new metaphysical claim that is to be taken seriously at time t should be motivated by, and only by, the service it would perform, if true, in showing how two or more specific scientific hypotheses, at least one of which is drawn from fundamental physics, jointly explain more than the sum of what is explained by the two hypotheses taken separately' (2007: 37).
52. For a classic definition of intrinsic properties see Langton and Lewis (1998), where the most general definition of it is that of a property possessed by its bearer independently of 'accompaniment' or 'loneliness'. OSR rejects any intrinsic property because there can be no *bearers* of properties to begin with.
53. While Derrida explained that 'the theme of *différance* is incompatible with the static, synchronic, taxonomic, ahistoric motifs in the concept of structure', he insisted, however, that such ideals of immutability do not uniquely characterise the concept of structure, since 'the production of differences, *différance*, is not astructural' but, conversely, it 'produces systematic and regulated transformations which are able, at a certain point, to leave room for a structural science' so that '*différance* even develops the most legitimate principled exigencies of "structuralism"' (1981: 27, 28).
54. Ladyman and Ross acknowledge our cognitive limitations when they write that '[w]e may not *be able to think* about structure without hypostatizing individuals as the bearers of structure, but it does not follow that the latter are ontologically fundamental' (2007: 155; emphasis added).

55. Ladyman and Ross, rejecting the idea of a fundamental level, explain that they renounce the very idea of hierarchical levels, since 'by "fundamental" physics we…refer to that part of physics about which measurements taken anywhere in the universe carry information' (2007: 55). But note McKenzie's (2012) important critique of OSR. Whilst rejecting the notion of a fundamental level, McKenzie disambiguates the at time inexplicit notion of fundamentality at work in the structuralist literature. He articulates two main ways to cash out fundamentality in terms of different kinds of priority relations between objects and structures: (1) relations of *supervenience*, where in his reckoning OSR fails, or (2) relations of *dependence* (modelled on Kit Fine's approach to ontological dependence), a most promising yet ultimately not totally adequate option for the OSRist. McKenzie opts for a moderate structural realism where structures are *no more nor less* fundamental than objects.
56. From his landmark paper 'Real Patterns' (Dennett 1991). John Haugeland (in Dahlbom 1993) offers an excellent exegesis of this paper, interpreted as a full-bloodied ontological essay.
57. In this section I want to pare down to the bone an extremely complex and dialectically articulated position that the authors define as Rainforest Realism (RR) (Ladyman and Ross: 2007: 233) (pitted against Quine's realism of desert landscapes), which, conjoined with the previously exposed OSR, informs an 'Information-Theoretic Structural Realism' (ITSR) (2007: 238). Without misrepresenting their ontology, I will leave aside some of the most technical details (and the entire re-framing of the concepts of causation and laws of nature in terms of recurrent structural relations among patterns) and, trying to streamline the articulated network of concepts buttressing ITSR, concentrate on those ontological elements that have a more direct bearing on my argument in this and the next chapter.
58. This ontological picture requires a revised vocabulary: the sciences do not study objects and their properties, but *track patterns* at different scales, from the microphysical to the socio-economical.
59. Frigg and Votsis conclude their review of the structuralist literature by noting that 'the jury still seems to be out on whether modern physics really favours OSR over its more traditional rivals. Moreover the question remains whether OSR, and ESR for that matter, can give an adequate account of the ontology and epistemology of other sciences' (2011: 269).
60. The debate on the causes and on the very existence of something called 'the Scientific Revolution' has been raging in the history of science for at least four decades (the literature is vast, but essential guideposts for the critical analysis of the concept of 'Scientific Revolution' are Cohen 1994; Hellyer 2003; Lindberg and Westman 1990; Osler 2000; Shapin 1996). However, the importance of the role played by the mathematisation of nature (a phrase originally coined by Koyré 1957) in shaping a more or less radical shift in the way we have approached and studied nature cannot be doubted. A recent compelling account is offered by Floris Cohen in his monumental work on the origin of modern science (Cohen 2010), reconstructing the historical trajectory of two distinct and long-standing approaches to the study of nature: the systematic and broadly philosophical approaches to the philosophy of nature (collectively labelled 'Athens') and the more disjointed set of abstract mathematical disciplines (labelled 'Alexandria').

61. This position only apparently resonates with Quine's well-known ontological reduction of physical objects to sets. Having noted how contemporary physics problematises the ontological notion of elementary particle, and having shown how 'spatio-temporally located physical objects' are empirically equivalent to 'spatio-temporal location' Quine goes on to argue that

    a further transfer of ontology suggests itself: we can drop the space-time regions in favor of the corresponding classes of quadruples of numbers according to an arbitrarily adopted system of coordinates. We are left with just the ontology of pure set theory, since the numbers and their quadruples can be modelled within it. There are no longer any physical objects to serve as individuals at the base of the hierarchy of classes, but there is no harm in that. It is common practice in set theory nowadays to start merely with the null class, form its unit class, and so on, thus generating an infinite lot of classes, from which all the usual luxuriance of further infinites can be generated. (1981: 17–18)

    However, Quine's reference to a determinate space-time location of objects is not supported by more up-to-date physics, and this is precisely the point that OSR wants to cash out. This, however, does not mean, as I will argue later, that a structuralist approach to physico-mathematical 'objects' cannot be compatible with set theory.
62. Michael Resnik, for example, has argued that

    because quantum particles are probability fields or something akin to them, they will impede attempts to distinguish between mathematical and physical objects on spatio-temporal or causal grounds; and... undetectable physical processes count against distinguishing mathematical and physical objects on the basis of physical detectability. Although the mathematical–physical distinction is central to the standard approach to the philosophy of mathematics, these claims underscore the need to clarify it and our current inability to do so... While we may find interesting epistemic differences between... structures, I see no reason for interesting epistemic differences to arise between the structures studied in physics and those studied in the traditional (non-global, non-foundational) branches of mathematics. (1990: 377)
63. Dedekind is often considered the founder of structuralism in mathematics, with particular reference to his classic essay *Was sind und was sollen die Zahlen?* ('What are and what is the meaning of Numbers?' in Ewald 1996). For a thorough examination of Dedekind's stance and his methodological and logical structuralism see Reck (2003).
64. See his famous remark to Frege that

    it is surely obvious that every theory is only a scaffolding or schema of concepts together with their necessary relations to one another, and that the basic elements can be thought of in any way one likes. If in speaking of my points I think of some system of things, e.g. the system: love, law, chimneysweep... and then assume all my axioms as relations between these things, then my propositions, e.g. Pythagoras' theorem, are also valid for these things. In other words: any theory can always be applied to infinitely many systems of basic elements. (In Frege 1980: 40–1)
65. As he argued, '[m]athematicians do not study objects, but the relations between objects; to them it is a matter of indifference if these objects are

replaced by others, provided that the relations do not change. Matter does not engage their attention, they are interested by form alone' (Poincaré 1905: 20). Poincaré can be singled out as being an influential source for both structuralism in the philosophy of science and the philosophy of mathematics.

66. The French collective's axiomatic method is founded upon the notion of mathematical structure. As they explain,

> [t]he common character of the different concepts designated by this generic name, is that they can be applied to sets of elements whose nature has not been specified; to define a structure, one takes as given one or several relations, into which these elements enter...then one postulates that the given relation, or relations, satisfy certain conditions (which are explicitly stated and which are the axioms of the structure under consideration). To set up the axiomatic theory of a given structure, amounts to the deduction of the logical consequences of the axioms of the structure, excluding every other hypothesis on the elements under consideration (in particular, every hypothesis as to their own nature). (1950: 225–6)

For a thorough and critical analysis of the validity of Bourbaki's concept of structure see Corry (1992).

67. The *locus classicus* for a concise exposition of Quine's generalised structuralism is Quine (1992).
68. Reck and Price list two principles guiding this methodology: '(i) What we usually do in mathematics (or, in any case, what we should do) is to study the structural features of [mathematical] entities. In other words, we study them as structures, or insofar as they are structures. (ii) At the same time, it is (or should be) of no real concern in mathematics what the intrinsic nature of these entities is, beyond their structural features' (2000: 345).
69. But see also Hellman (1989) and Parsons (2004, 2008) for alternative defences of structuralism.
70. Shapiro (1997) and Resnik (1997). These build upon work done in the previous decades: see Resnik (1981, 1982) and Shapiro (1983).
71. In a landmark paper Benacerraf (1965) argued that there is no way to adjudicate between different set-theoretic reductions of the natural number sequence. In particular, he highlights the underdetermination number system of two young fictional characters 'Ernie' and 'Johnny', budding mathematicians trained to identify numbers with, respectively, the incompatible set theoretic approaches of Ernst Zermelo and John Von Neumann (isomorphic yet consisting of different elements). Benacerraf concludes that not only are numbers not sets, but 'numbers are not objects at all, because in giving the properties (that is, necessary and sufficient) of numbers you merely characterize an abstract structure – and the distinction lies in the fact that the "elements" of the structure have no properties other than those relating them to other "elements" of the same structure' (1965: 70). Together with his later epistemological worries about the knowability of causally inert objects (in Benacerraf 1973; see my discussion in section 2.4), this conclusion led Benacerraf to deny ontological reality to numbers. *Methodological* structuralism does not entail *ontological* structuralism.
72. But note that the inverse does not hold: van Fraassen wryly notes that '[o]ne of my great regrets in life is that I do not have a philosophy of mathematics' (in Stapleton 1994: 269).

73. Cf. Badiou's 'anti-object' rejection of Leibniz's PII (in Hallward and Peden 2012: 165–6). Shapiro explicitly rejects the PII in MacBride (2006).
74. Resnik explains that '[m]athematical objects are incomplete in the sense that we have no answers within or without mathematics to questions of whether the objects one mathematical theory discusses are identical to those another treats; whether, for example, geometrical points are real numbers' (1997: 90).
75. From both the mathematician and the systems which instantiate it.
76. Hellman refers to this position, referring to abstract structures generalizing abstract mathematical objects, as 'hyperplatonist' (in Shapiro 2005: 542). Dummett dismissed a (Dedekind-inspired) version of structuralism – committed to the existence of abstract structures – as 'mystical' (1991: 295).
77. After Parsons' reading (1990) of Dedekind's and Benacerraf's position. The label most straightforwardly applies to Hellman's *in re* variety of modal structuralism, where mathematical structures are seen as describing with what *would be the case* in any system of objects satisfying their axioms (see Hellman 1989).
78. MSists offer an account of the knowledge of structures which resembles Maddy's own perception-based approach (as well as Kitcher's), arguing that we can achieve knowledge of mathematical patterns from experience of physical instantiations. Developing this insight of concrete pattern recognition, Resnik argues that 'if the point of positing mathematical objects is to describe certain patterns, then it is plausible to allow that systems of physical objects instantiating these patterns can inform us of properties of mathematical objects' (1997: 224). He then proceeds, in a Quinean-holistic way, to postulate the indispensability of these mathematical objects. Shapiro's epistemology, on the other hand, relies on a process of linguistic abstraction and on the idea that a successful implicit definition (the characterisation of a number of items in terms of their relations to each other) fully characterises a structure or a system and that '[s]tructuring a chunk of space-time into a system of objects involves mobilizing linguistic and other conceptual machinery [that of implicit definition]' and that 'this mobilization is all that is needed to formulate and discuss structures as such' (1997: 257). Both these positions, like Maddy's own, place the issue of the knowledge of mathematical entities on a continuum with that of the knowledge of physical ones, and thus have the general consequence of blurring the abstract–concrete boundary.
79. Note that it is debatable whether Benacerraf is to be read straightforwardly as offering a structuralist *solution* to his dilemma. Paseau comments that 'though his paper inspired later structuralisms, it is hard to say with confidence that structural reductionism is Benacerraf's preferred way out' (in Bueno and Linnebo 2009: 52).
80. Nancy Cartwright and Mark Wilson have argued, in different ways, against the ability of mathematical descriptions to fully match the physical systems they describe without the employment of practical shortcuts, like abstracting away 'unruly' elements unable to be fully subsumed under a rigid mathematical mapping or the application of a patchwork of ad-hoc mathematical techniques. However, while Cartwright draws profound ontological conclusions from the stratagems of applied mathematics, ultimately arguing for the

anti-foundational worldview of a dappled, non-unified world (1983, 1999), Wilson, adopting what Deleuze would call a 'problematic' approach to mathematics, argues for the adoption of a form of 'mathematical opportunism', rejecting the optimist view of a mathematics that captures physical reality without reminder but eschewing anti-realist conclusions. Wilson argues that '*mathematical opportunism* is based upon the belief that the successes of applied mathematics require some alien element that cannot be regarded as invariably present in the physical world' while insisting that 'there is nothing in this thesis that should tilt us towards "anti-realism" about anything: it simply states that the processes upon which practical mathematics is based might prove too constrained to suit all of the circumstances of the physical world snugly' (2000: 297, 299).

81. Hence my endorsement of OSR, a position grounded on results coming from the physical sciences; perhaps not uncontroversial but certainly worth serious consideration. In other words, I am not primarily concerned with the manifest conformity of the behaviour of physical systems with mathematisable relations but with the unavoidable employment of mathematical means to describe the structural properties of unobservable 'entities' at elementary scales. Of course, the 'real patterns' thesis extends the naturalist approach to larger-scale patterns, but without having these patterns instantiate a universal Pattern. In Ladyman and Ross's extended, scale-relative ontology patterns enjoy an ontological independence that particulars-instantiating-universals do not.
82. Kantorovich (2009) considers the post-1960s classification of elementary particles in representations of internal symmetry groups to argue for a 'weak' form of OSR, where such symmetry structures are the deepest layer in the ontological hierarchy, such that 'there is a physically possible world where the instantiations of the structure do not exist' (2009: 81).
83. This is not incompatible with the metaphysics of OSR. As Ross and Ladyman themselves clarify: 'there is the well-known concern that adopting a set-theoretic framework compromises the underlying metaphysics of OSR, since it suggests a commitment to sets of objects. One response is to insist on the 'thin' notion of object... and argue that such commitment does not take us beyond the bounds of structuralism, since these objects are themselves individuated structurally' (in Bokulich and Bokulich 2011: 32). As they further explain: 'we are basically saying the following: although the set-theoretic representation of structure, <A, R> (where A is a set of elements and R a family of relations) must be written down from left to right, from an ontological perspective it should be read semitically that is, from right to left' (in Bokulich and Bokulich 2011: 33). The set-theoretic ontology the eliminative structuralist requires is made of non-structurally *existing* sets, not by non-structurally *identifiable* ones. Set theory merely offers, so to speak, the rough material (an empty set devoid of identifiable individual properties and self-existence) the qualification of which is given only and thoroughly via structural relations. As Shapiro acknowledges '[o]n all versions of structuralism, the nature of the objects in the places of a structure does not matter – only the relations among the objects are significant. On the ontological [eliminativist] option, then, the only relevant feature of the background ontology is its *size*' (1997: 88).

84. The rejection of the *ante rem* structuralist's ontology of universals does not entail an exclusion of their *epistemological* reflections. As I argued in Chapter 2 vis-à-vis Maddy's work, accounts of the genesis of our mathematical perceptions are precious resources in the project to naturalise, bottom-up, the development of our mathematical concepts. The epistemological ladder need not be climbed all the way to universal abstract structures.

5 Truth and Randomness

1. Brassier warns against confusing Heidegger's and Badiou's accounts by explaining that '[w]hereas Heidegger's "Being" is in transcendent exception to everything that is because it is *more than* anything, Badiou's void is in immanent subtraction to everything that is because it is *less than* anything' (in Hallward 2004: 243). The difference is clear enough, but does not offer much comfort to the naturalist's scepticism towards the very idea of 'Being' as *distinct* from the 'everything' of nature.
2. That is, left unresolved, as I noted in my Introduction, by the convoluted account offered in *Logics of Worlds*, where it remains unclear what the connection between objective appearances and things in their ontological reality is, and according to which causal processes the impersonal 'transcendental legislation' can come to have any organising force over phenomenal reality.
3. See Badiou (2006d: 40).
4. Consider, for example, how Badiou suggests the distinction between mere knowledge and novel truth to be isomorphic with Heidegger's 'distinction between truth, *aletheia*, and cognition or science, *techne*' (2001b: 249).
5. As Hallward notes '[i]t is not clear… that a radical (and materialist) univocity can survive this dualism [of a thought of Being utterly different from the ontic multiplicity of beings]' (2003: 276). Note that the matherialist/structuralist stance, which I prefer to Badiou's own, cannot be accused of erasing the ontological difference, since the latter is simply recast in fully immanent terms: structure is indeed different from structur*ed*, yet only *formally*. A structure is not a 'highest Being' over and above structured entities/patterns, nor the *genus* of which the latter are species, but the immanent domain of relations out of which patterns are individuated.
6. See Heidegger (1982: 17) for an early formulation.
7. Importantly, Badiou 'willingly admit[s] that absolutely originary thought [of Being] occurs in poetics and in the letting-be of appearance' (2006a: 125), yet immediately adds that only after the Greek mathematical 'event' could ontology proper be articulated. Of course, it cannot be denied that poetic expression has been one of the first ways in which, historically, humans have attempted to come to grips with, and to make sense of, reality. But the assumption that, while the poetic expressive strategy was mistaken and laden with fallacious metaphysical prejudices, the 'Being' referred to by Parmenides (as well as the roughly equivalent fundamental principles of reality as articulated by other philosophical traditions) was a legitimate object of inquiry is where Badiou and I part ways. No matter how 'subtractive' our approach to it might be, how primordially multiple as opposed to unitary we take this

principle to be, the adoption of a scientific conceptual framework compels us to abandon any reference to a supreme principle of or ground for the multifarious appearances of reality. Note: I am *not* dismissing inquiry into the limits of the sayable, pushing sense beyond its boundaries; I am rejecting the hypostatization (even when self-effacing) of that no-thing which allegedly lies in this beyond, making transgression possible. Transgression is an issue for semantics, not metaphysics.

8. Very similar objections towards Heidegger's methodology have been raised before in analytic quarters. As early as 1929, Gilbert Ryle, in his review of *Being and Time*, lamented that 'certain theories of human nature have been interpolated into Heidegger's analysis' and that 'an anthropologistic Metaphysic [sic] seems to have been read out of our everyday experience' (in Murray 1978: 61).
9. Note Badiou's peculiar placement in this tradition: whilst self-avowedly faithful to Plato, Badiou hardly acknowledges how his interest in offering a better answer to Heidegger's ontological questions places him in the *Aristotelian* tradition running from the Stagirite to Heidegger through, at least, Aquinas, Scotus and Brentano. Badiou, in other words, attempts to respond to a question that is Aristotelian in formulation and pedigree (that of Being *qua* Being) with a Platonic answer drawn from his reading of the *Parmenides*.
10. I am not adopting the anti-metaphysical Carnapian stance (accusing the ontologist to unwisely inflate contingent quirks of our employment of language) which I criticised in the previous chapter. My position is buttressed by the conviction that, *at this level of generality*, there is nothing meaningful that can be said when one systematically excludes empirical considerations. Ontology *can* be done, but only when proceeding hand in hand with the empirical sciences. It is one thing to undertake an *a priori* transcendental inquiry on the conditions of possibility of our empirical contact with the universe (a project I endorse, in both its Kantian-conceptual and Bhaskarian-realist forms), but it is another to indulge in an utterly general speculation on a 'Being' conceived as the allegedly featureless condition of possibility for *all* that there is.
11. I should mention that, haughty dismissal by the OSRists notwithstanding, the field of neo-Aristotelian metaphysics which is much more sympathetic to the notion of 'substance' and 'powers' is alive and well, and staunchly defended by able philosophers. For recent examples, see Mumford and Tugby (2013) and Tahko (2012).
12. Recall that, for Aristotle, there are three theoretical philosophies: mathematics, natural science and theology (metaphysics): 'natural science deals with things which are inseparable from matter but not immovable, and some parts of mathematics deal with things which are immovable, but probably not separable, but embodied in matter; while the first science deals with things which are both separable and immovable' (*Meta*. E.1, 1026a, 15–18). The structuralist rejects the intelligibility of the 'separability' and 'immovability' criteria as applied to mathematics, and considers the latter as replacing both natural science and theology.
13. Varzi clarifies the 'inventory' approach to ontology, explaining that 'it is evident that the inventory activity which the ontologist takes as her task

should not be intended as a census of all the citizens of this world: it is rather a matter of compiling a list of all kinds of entities to include in the catalogue' (2005: 12; my translation). Note that my kind of structuralist approach undermines the talk of *kinds*. If it even makes sense to say so, one should claim (as I have argued in the previous chapter) that for the math-erialist structuralist there is but *one* kind of entity: the empty set *qua* minimal ontological commitment, defying concrete/abstract distinctions. The variety of the universe, at different scales, can be cashed out with talk of patterns.

14. I borrow this term from Humphreys (in Ladyman et al. 2013).
15. Badiou argues that 'in *non-ontological* situations, foundation via the Void is impossible. Only *mathematical ontology* admits the thought of the suture to being under the mark Ø' (2006a: 188). But, since this argument is grounded on the separation between the ontological and the ontic and the split between natural and historical situations that I have already rejected (Badiou clearly explains that '[r]ecourse to the void is what institutes, in the thought of the nature/history couple, an *ontico-ontological difference*' (ibid.)), such prohibition does not bear any weight on my argument.
16. Incidentally, this could allow us to defend a realist interpretation of Derrida as well: if any structural system, be it linguistic or mathematical (see Derrida's remarks on mathematics in 1981: 34, 35), immanently deploys and transcends its own limits, and if reality itself can be ultimately described in purely structural terms (a system of relations without positive terms), then Derrida's unfolding the aporiae of structuralism can be generalised into an inquiry into features of reality, and the paradoxical status of *différance* does not merely affect linguistic presentation, but everything that is *qua* structured. (Cf. Floridi's adoption of difference as a basic ontological principle, exposed in Section 4.3.)
17. The term was famously employed by John Archibald Wheeler to describe the kind of universe entailed by his 'it from bit' motto (Wheeler in Hey 1999): Wheeler's informational universe required the intervention of experimenting subjects to compel reality to 'respond' and turn information into physical matter. Using the example of the experimental detection of a photon, Wheeler writes that '[w]e know perfectly well that, the photon existed neither before the emission nor after the detection. However, we also have to recognize that, any talk of the photon "existing" during the intermediate period is only a blown-up version of the raw fact, a count' (in Hey 1999: 311). The employment of the term 'count' should alert us: even if Badiou defends his notion of count as non-subjective, and suggests (in *Logics of Worlds*) a transcendental 'without a subject', these somewhat arcane forms of '(non)subjective' constitutive intervention onto the world are a concern to the naturalist, and remain unclearly distinguished from positions like Wheeler's.
18. See my discussion of Maddy, Kitcher, Dehaene, Changeux and Carey in Chapter 3.
19. Cf. '[t]he sufficient reason, therefore, which needs no further reason; must be outside of this series of contingent things and is found in a substance which is the cause of this series or which is a necessary being bearing the reason for its existence within itself; otherwise we should not yet have a sufficient reason with which to stop. This final reason for things is called *God*' (Leibniz 1989: 639).

20. I owe this insight to Pete Wolfendale. Although the concept of limit was already, if imperfectly, employed by Newton, the rigorous definition of the modern mathematical notion of limit is generally attributed to Cauchy's work on infinitesimal calculus, in the 1820s. See Grabiner (1981).
21. I implicitly refer to Peirce's truth as ideal limit of scientific inquiry (Peirce himself adopts the mathematical example of π as 'an ideal limit to which no numerical expression can be perfectly true' (*CP*: 5.565)) and to Sellars's rejection of scientific knowledge as a *static* achievement, and rather seen as 'a self-correcting enterprise which can put *any* claim in jeopardy though not *all* at once' (1963: 170). Of course, Sellars's realism led him to criticise Peirce's lack of an 'Archimedean point outside the series of actual and possible beliefs in terms of which to define the ideal or limit to which members of this series might approximate' (1968: 142); yet he agreed with Peirce's *regulative ideal* of an absolutely adequate conceptual framework motivating inquiry (and affording a perfect fusion of manifest and scientific images) but never fully accomplished.
22. For a defence of such a Wittgensteinian interpretation of the infinite as *shown* to us but never *known* by us, see Moore (1990).
23. See Chomsky (2007, Part II, Chapter 4).
24. Price refers to a 'subject naturalism', according to which 'philosophy needs to begin with what science tells us *about ourselves*. Science tells us that we humans are natural creatures, and if the claims and ambitions of philosophy conflict with this view, then philosophy needs to give way' (in De Caro and Macarthur 2004: 73).
25. Johnston (2008b) most perspicuously highlights the long shadow of Kant in Badiou's work.
26. Hence my employment, in what follows, of themes drawn from Peirce, that 'gifted composer of variations on Kantian themes' as Sellars (1968: vii) memorably put it.
27. Fraser notes that '[t]he word 'void' is surprisingly equivocal in Badiou's writings' and that 'we can discern, in Badiou's work, at least four distinct senses of "void"' which 'are not always easy to untangle' (in Corcoran forthcoming). What is important to highlight in this context is how the set-theoretical basic regress-stopping 'atom' void/empty set is *neither* the same 'void' synonymous with inconsistent multiplicity/Being nor that 'void' which Badiou will address as lying between presentation and representation.
28. Badiou interrogates the concept of nature through this conceptual lens, asking whether a pertinent concept of nature can be formulated within an ontology of multiplicity. He notes that '[p]aradoxically, it is again Heidegger who is able to guide us here' (2006a: 127). Borrowing Heidegger's insights on nature *qua* consistent maintenance-in-the-open Badiou concludes that the ontological form of nature is the natural multiple, the set of all ordinals, presenting 'a particular consistency, a specific manner of holding-together' (2006a: 127). It is a short step for him to then claim that 'Nature does not exist' (2006a: 140) since there cannot be a consistent, non-paradoxical set of all sets. Hence Badiou's rejection of Spinozism: for Spinoza, the multiplicity of Being is always recaptured by a divine metastructure, a *Deus sive Natura* wherein presentation and representation eternally coincide. I will argue

below that to defend inconsistent nature does not require an abandonment of naturalist immanentism.

29. Once again, Badiou makes explicit both his debt to Heidegger and their shared scorn of 'Nature' as a locus of truth, clarifying that he 'retain[s] from Heidegger the germ of his proposition: that the place of thought of that-which-is-not-being is the non-natural; that which is presented other than natural or stable or normal multiplicities. The place of the other-than-being is the abnormal, the instable, the antinatural' (2006a: 174).
30. The Axiom of Foundation states that every set is grounded, and it makes possible the ordered hierarchy of the set-theoretical universe. Potter argues that 'when the axiom of foundation was first mooted, it was regarded more as a convenient tool for metamathematical reasoning than as an evident truth' (2004: 295). Badiou explains that this axiom '*forecloses extraordinary sets* [sets belonging to themselves] *from any existence, and ruins the possibility of naming a multiple-being of the event*' and so '[O]ntology declares that the event is not' (2006a: 190).
31. Denunciation of Heidegger's own poetic ontology notwithstanding, Badiou is all too eager to rehabilitate poetry when the event breaks the ontological order. So, 'the nomination of an event – in the sense in which I speak of it, that is, an undecidable supplementation which must be named to occur for a being-faithful, thus for a truth – *this nomination is always poetic*. To name a supplement, a chance, an incalculable, one must draw from the void of sense, in default of established significations, to the peril of language. One must therefore poeticize, and the poetic name of the event is what throws us outside of ourselves, through the flaming ring of predictions.' (2004: 100; emphasis added)
32. Badiou defines a subject of truth 'a body behaving as the support of a formalism' (2009: 69).
33. Cf. '[d]eductions are truth preserving ['veridical moves within the situation', as Badiou would have it], whereas successful abductions may be said to be *truth producing*. This ampliative reasoning is sometimes done by *introducing new vocabulary in the conclusion*' (Josephson and Josephson 1994: 13; emphasis added).
34. As Badiou's 'master' Lacan has it: '[t]ruth is nothing other than that which knowledge can apprehend as knowledge only by setting its ignorance to work' (2002: 675).
35. If Plato's *Parmenides* is Badiou's starting point for his ontology, the *Meno* is the principal Platonic inspiration for his reflections on truth, a dialogue pivoting (through the celebrated set piece of the boy slave) on the paradox of the attainment of *a priori* (mathematical) truth about what one does not yet know.
36. Riffing on the traditional definition (side-stepping Gettier), for Badiou knowledge is *merely* justified belief, while 'truth' is indeed the process of holding a belief against the odds of justification, or diachronically with an infinite process of justification.
37. In technoscientific contexts, this idea of a heuristically productive obstacle is well articulated by Edward Constant's notion of 'presumptive anomaly': when a given applied science is able to foresee 'that under some future condi-

tions the conventional system will fail (or function badly) or that a radically different system will do a much better job' (1980: 15).

38. For a full-blooded realist theory of truth see Devitt (1997). Metaphysics-light minimalist theories or deflationary ones rejecting the idea of a 'nature' of truth are offered by Alston (1996) and Horwich (2004). Paradigmatic exemplars of the anti-realist stance are Dummett (1978) and, in another context, van Fraassen (1980).
39. Note the obvious analogy with philosophy of mathematics: as in ontological matters Badiou's stance is neither an intuitionism nor a Platonism, so his position on truth is neither a constructivism nor an absolutism.
40. So Badiou's concept of truth (or perhaps better: Badiou's *operational elucidation* of such concept) echoes Deleuze's claim that '[i]n every respect, truth is a matter of production, not of adequation. It is a matter of genitality, not of innateness or reminiscence' (2001: 154). Gutting has correctly claimed that we can identify 'a distinctive unity in French philosophy since 1960: a fundamental concern with thinking what is conceptually impossible' (2011: 184), and, reflecting on the same historical period, Badiou indeed explains that '[w]e wanted something quite unusual, and admittedly problematic: our desire was to be adventurers of the concept' (2005b: 77). An important caveat is, however, that unlike a great many of his contemporaries, Foucault first among them, for Badiou the endogenetic nature of truth does not make it 'a thing of this world' (see Foucault 1980: 131): the singularity of his account conspicuously lies in the demonstration of how an entirely immanent (and *collective*) process can ultimately deliver a truth transcending the boundaries of epistemic consensus.
41. The stance of epistemic infinitism, often attributed to Peirce (see Aikin 2009), and recently witnessing a certain resurgence (see Aikin 2007 and Klein 1998, 2006), holds – against foundationalism and coherentism – that an infinite series of inferential relations are necessary for epistemic justification, thus side-stepping the epistemic regress problem. Peter Klein explains how there is 'an inherent, non-dogmatic tendency in infinitism. For the infinitist grants that she has not finished the process of justifying her beliefs. There is always a further step that can be taken should we become dissatisfied with the point at which we stopped the progress of inquiry' (2006: 10).
42. I stress the formal, not *moral*, nature of those demands. Simmons and Aikins (2012) have proposed an interesting reading of Levinas as endorsing epistemic infinitism by having Levinasian ethics ground a Clifford-inspired ethics of belief. Their reconstruction insightfully demonstrates a structural similarity between Levinas's ethics of infinite transcendence and the endless requests for justification of infinitary epistemology. The adoption by Badiou of the same epistemic process, however, is not motivated by a *transcendent* (and therefore ethical) demand. Epistemological norms are not, as they are for Levinas, grounded on ethics as first philosophy. Keeping firmly in place Clifford's admonitions to the effect that committed evidentialism is a moral imperative, it seems to me that to defend the Levinasian project as epistemically workable is in explicit contravention of those Enlightenment values of *autonomous* responsibility and conformity to *formal* rules of reason I highlighted in my Introduction.

43. Via the addition of the Axiom of Constructibility V = L to the ZF axioms: having equated the universe of sets V with the constructible one L (a universe where each set is composed from the empty set in a definable series of explicit steps), CH can be proved to be veridical in L.
44. This does not imply that the distinction between situation and state need be abandoned. However, in an epistemological register, it should be recast in social terms: taking a scientific situation as an example, the re-securing of the situation by the state can be reinterpreted as the collective action of a given scientific community upon itself to shield its conceptual nodes from unwelcome revolutions. Note that this example is not far from Badiou's own explications of the situation/state dialectic, most often resorting to political metaphors: strikingly, however, this example is much *more* intelligible in my non-ontological rephrasing than in Badiou's own, where the actual link between the formal laws of Being described by set theory and concrete political situations is never fully clarified.
45. The work of both Nancy Nersessian (2008) and Susan Carey (2009) are magisterial examples of the kind of cognitive-historical analysis aimed at elucidating the actual *process* of conceptual change in both everyday and scientific practice.
46. Which is not to say that it is a *necessary* moment; human agents are *not* systematically rational. I am dealing here with scientific thought as the most refined and least error-prone (but not incorrigible) form of human intellectual activity. Rational thought does not come cheap, but it would be a mistake to call 'rational' only *perfect* inferential practices. Rational thought is an endlessly self-correcting enterprise.
47. Little naturalist elucidation on the genesis of subjectivity is offered by Badiou: its pre-evental existence problematically denied, its advent on the scene strictly dependent on the discernment of an event, and its actions stringently bound to the pursuit of the event's consequences: the subject's 'entire being resides in supporting the realization of truth' (2006a: 397).
48. Badiou's protestations to the contrary notwithstanding, it is hard *not* to find Badiou's example of St Paul (2003a) as exemplar subject of evental truth, revealing. Phelps, expanding Bosteels' suggestion that there is an 'antiphilosophical temptation at work in Badiou's philosophy' (2008: 179) argues that indeed 'the heart of Badiou's philosophy appears determined by an internal anti-philosophical element, an element that is inseparable from theology... Given the fact that the anti-philosopher institutes a discourse of rupture, I would suggest that it is a necessity, given Badiou's understanding of the advent of truth as evental' (2013: 159) and contends (persuasively, it seems to me) that Badiou's ideal of heroically militant subject 'dovetails with the Christian conception of the saint' (2013: 159).
49. See Sellars (1963: 100).
50. As Brassier puts it:

    [t]o acknowledge a constitutive link between subjectivity and rationality is not to preclude the possibility of rationally investigating the biological roots of subjectivity. Indeed, maintaining the integrity of rationality arguably obliges us to examine its material basis. Philosophers seeking to uphold the privileges of rationality cannot but acknowledge the cogni-

> tive authority of the empirical science that is perhaps its most impressive offspring. (2011b: 9)

51. Paul Humphreys writes that

> [w]hat constitutes a pattern for a human is more restricted than (or is just different from) what constitutes a pattern for a computer. This suggests that philosophers need to do one of three things; expand their repertoire of representations to include those used by computational devices and instruments, despite the fact that those representations do not conform to any that are intuitively accessible to humans; concede that certain current and future scientific activities lie beyond human understanding...a position that is at odds with the fact that we can understand at least some of these activities; or push their explorations of sub-conceptual, non-representational, and non-conceptual content beyond the realms currently explored in artificial intelligence, such as the use of dynamical systems theory or neural net representations. The second and third options entail that there are parts of ontology that lie beyond what is currently intuitively accessible, while the first option is unlikely to be successful without a considerable amount of conceptual retraining. (In Ladyman et al. 2013: 65–6)

52. See, for example, Langley et al. (1987); Boden (2004); Magnani (2001, 2009); Magnani and Li (2007). That we haven't yet achieved to create a fully 'creative' computational system has arguably more to do with the limits of current paradigms in AGI research than it does with physical and logical possibilities. David Deutsch (2012: np) argues that 'unlike any functionality that has ever been programmed to date, [creativity] can be achieved neither by a specification nor a test of the outputs. What is needed is nothing less than a breakthrough in philosophy, a new epistemological theory that explains *how* brains create explanatory knowledge and hence defines, in principle, without ever running them as programs, which *algorithms* possess that functionality and which do not'. Boden displays a more cautious optimism, claiming that 'the question of whether a computer could ever "really" be creative is currently unanswerable, because it involves several highly contentious philosophical questions' (in Paul and Kaufman 2014: 242).
53. See particularly Turing's seminal (1936) paper on computable numbers, aimed at proving that there is no general decision procedure for first-order logic, which demonstrated the power *and* the limits of 'computers' and introduced the notion of what became known as a universal Turing machine.
54. The thesis today known as the Turing–Church thesis implies that the computational capabilities of a Turing machine (or indeed of any other formal model of computation) are universal computational limits, and thus the same as those of a human 'computer' carrying out mechanical algorithms: the intrinsic *limits* of a Turing machine (identified in the so-called 'halting problem', or the impossibility for a given computer to establish whether another computer instantiating a program will ever come to a halt and deliver a definite output) are *also* limits for human reasoning. Turing himself was rather cautious regarding the consequences of this thesis (but see Turing 1950 for a pioneering discussion of 'thinking machines', arguing against skeptical arguments questioning a machine's ability to fully replicate human cognitive abilities) and it was only several years later the Turing–Church thesis was explicitly applied to buttress the equivalence of human brains

with Turing machines made neural circuitry propagating digital information (by the work of McCulloch and Pitts to begin with and Weiner and Von Neumann later) thus ushering in the era of artificial intelligence research.

55. See Chaitin (1975, 1987, 1998, 1999, 2007).
56. In their classic *Principles of Mathematical Logic*, Hilbert and Ackermann explain that 'a general method of decision would consist of a certain recursive procedure for the individual formulas which would finally yield for each formula the value truth or the value falsehood' but note that '[r]esults by A. Church [the authors were still not familiar with Turing's paper] based on papers by K. Gödel show that the quest for a general solution of the decision problem must be regarded as hopeless' (1950: 124).
57. The understanding of scientific theorising as the search for the simplest program/theory capable of producing/explaining the data (seen as a string of binary digits) is due to Ray Solomonoff's (1964a, 1964b) formal theory of induction.
58. This definition of a measure of the computational resources needed to specify a given string of information is known as Kolmogorov–Chaitin complexity, the two mathematicians having independently reached this insight in the 1960s (together with Solomonoff).
59. Note that Chaitin here explicitly refers to Leibniz's PSR, his irreducibly random numbers proving it to be wrong. (Chaitin assigns to Steven Wolfram the role of a contemporary Leibnizian committed to a thoroughly rational and deterministic universe, where all randomness is only pseudo-randomness. See Wolfram 2002.) Leibniz can be seen to be in agreement with the definition of randomness as a function of irreducible complexity offered by Chaitin since, for Leibniz, 'God has chosen that world which is the most perfect, that is to say, which is at the same time the simplest in its hypotheses and the richest in phenomena' (1989: 306); in other words, simplicity is maximal output from minimal initial information (fed in by God). The difference is, of course, that whilst for Leibniz *all there is* is intrinsically ordered (by God), for Chaitin only a vanishingly small fraction of the universe fits the definition of non-random.
60. This position was introduced by thinkers like J.A. Wheeler and Konrad Zuse and is today endorsed by partisans of digital physics. See Zenil (2012) for a collection of essays on the 'computable universe', including a re-edition of Zuse's (1969) seminal essay 'Calculating Space'. It does not only apply to the physical: Chaitin's latest project (2012), arising from the insight that DNA can be seen as a program to produce various species as outputs, is to offer a computational approach to biology (a 'metabiology'), thus attempting to develop biological theorems.

Conclusion

1. For a quick overview of the Quine–Sellars debate on this topic see O'Shea (2007: 73–6).
2. Dilpert's graph-theoretical stance has recently enjoyed something of a revival through Alexander Bird's dispositional monism of relationally individuated pure powers (Bird 2007) and in the discussion between David Oderberg (2011,

2012) and Nick Shackel (2011) on the very plausibility of graphical structuralism, in both Dilpert's original form and Bird's reworking of it. Ladyman (2007) has also referred to graph theory in order to defend the contextual individuation of 'entities' implied by OSR. See Chakravartty (in Landry and Rickles 2012) for a critique of the cogency of the graph theoretical approach to elucidate the structure of physical reality: Chakravartty is not willing to give up the abstract/concrete distinction, to him an indefensible 'Platonist or Pythagorean extremism' (in Landry and Rickles 2012: 199).

3. See also his previous defence of 'mathematical fundamentalism' in Hut et al. (2006).
4. See Scheibe in Echeverria et al. (1992) for a description of this problem.
5. A difference between me and Tegmark is that while he is 'convinced that there is a deep explanation for it all' and that 'the only question [is] whether we can understand it from the frog perspective of our limited human minds' (in Hut et al. 2006: 783), I am highly sceptical about this commitment to the rational organisation of the universe *pre-existing* our rational-epistemic efforts, even while agreeing that there are limits to our cognitive capacities (see Chapter 5).
6. This line of inquiry is best presented in the exchange between Michael Friedman and Mark Wilson (based on the arguments of their respective masterworks, Friedman 2001 and Wilson 2006) regarding the evolution of conceptual frameworks and the consequences of a 'return to Kant' in contemporary philosophy of science (see their entries in Ladyman et al. 2013). It seems that whatever philosophical progress lies ahead of us, it will have to go through Kant (and a Kantian engagement with contemporary science) rather than side-step him.
7. In this regard, I need to stress again how the matherialist stance upheld here, preserving the tangled dualism of manifest and scientific image, should not be interpreted as a reductionism or an eliminativism. Whilst within the scientific image description and explanation are inseparable, *within the manifest image* the register of ontological description is independent from that of explanation: there might be facts that cannot be couched in a scientific vocabulary (hence the mutual dependency of the two images: the motive for the pursuit of a *scientific* understanding lies in the space of reasons of the *manifest* image of persons and their goals). At the same time, the scale-relativity of the pattern worldview allows us to smooth over the gap between the two images, and preserve the relative independence of inquiries into facets of reality of higher complexity, from the non-physical sciences (especially the life sciences) to phenomenology, whilst offering an information-theoretic meta-framework capable of subsuming reference to both subatomic particles and sentient organisms.
8. My apparent agreement with Rortyan pragmatism does not go hand in hand with his relinquishment of a truer language. Rorty, his 'veneration for Wilfrid Sellars' notwithstanding, invites us to abandon the quest for 'a unified picture, and for a master vocabulary' (in De Caro and Macarthur 2010: 58). But I think that this is a simplification of the *regulative ideal* of a synoptic fusion of images (vocabularies/conceptual frameworks) that *must* be preserved as an always receding target: navigating between the scientistic hubris of total understanding and ironic-quietist abandonment of the epistemic project.

My worldview calls for an alethic pluralism: the equation of mathematical and physical structure vindicates, and makes fully intelligible (no mean feat given the notorious problems with this stance), the *correspondence* view of truth, to the extent that for a mathematical statement to 'correspond' to a physical reality literally means for the former to be perfectly isomorphic with the latter (ultimate truth) – mathematical signs directly *picture* physical reality. On the other hand, at macroscopic scales we should loosen the correspondence constraints and accept less demanding but explanatorily more useful notions of truth (perhaps along the lines of semantic assertability) capable of accounting for the ambiguity engendered by evolving conceptual frameworks (conventional truth). The 'reimagined' Badiouian concept of truth I explained in Chapter 5 cuts across this dualism: more the description of a scalable process than a definition, the abductive pursuit of truth can be applied to both the formal and the macroscopic level.

Bibliography

Aikin, Scott F. (2007). 'Meta-epistemology and the Varieties of Epistemic Infinitism'. *Synthese*, 163(2): 175–85.

Aikin, Scott F. (2009). 'Prospects for Peircean Epistemic Infinitism'. *Contemporary Pragmatism*, 6(2): 71–87.

Ainsworth, Peter M. (2010). 'What Is Ontic Structural Realism?' *Studies in History and Philosophy of Modern Physics*, 41(1): 50–7.

Aliseda, Atocha (2006). *Abductive Reasoning: Logical Investigations into Discovery and Explanation*. Dordrecht: Springer.

Allen, Amy (2003). 'Foucault and Enlightenment: A Critical Reappraisal'. *Constellations*, 10(2): 180–98.

Allison, Henry (2004) [1983]. *Kant's Transcendental Idealism: An Interpretation and Defense*. New Haven, CT: Yale University Press.

Alston, William P. (1996). *A Realist Conception of Truth*. Ithaca, NY: Cornell University Press.

Althusser, Louis (1990). *Philosophy and the Spontaneous Philosophy of the Scientists and Other Essays*. London: Verso.

Althusser, Louis (2006). *Philosophy of the Encounter. Later Writings, 1978–87*. London: Verso.

Antonini, Aldo and Altamore, Giovanni (eds) (2010). *Galileo and the Renaissance Scientific Discourse*. Rome: Edizioni Nuova Cultura.

Ayer, A.J. (1971). *Language, Truth and Logic*. London: Penguin.

Bachelard, Gaston (1963). *Le Matérialisme Rationnel*. Paris: Presses Universitaires de France.

Bachelard, Gaston (1972). *L'Engagement Rationaliste*. Paris: Presses Universitaires de France.

Bachelard, Gaston (1984) [1934]. *The New Scientific Spirit*. Boston: Beacon Press.

Bachelard, Gaston (2002) [1938]. *The Formation of the Scientific Mind: A Contribution to a Psychoanalysis of Objective Knowledge*. Manchester: Clinamen Press.

Badiou, Alain (1994). 'Being by Numbers: Interview with Artists and Philosopher Alain Badiou'. *Art Forum*, 33(2): 84–7, 118, 123–4.

Badiou, Alain (1998). 'Politics and Philosophy' [interview with Peter Hallward]. *Angelaki* 3(3): 113–33.

Badiou, Alain (1999). *Manifesto for Philosophy*. Albany, NY: SUNY Press.

Badiou, Alain (2000a). *Deleuze: The Clamor of Being*. Minneapolis, MN: University of Minnesota Press.

Badiou, Alain (2000b). 'Metaphysics and the Critique of Metaphysics'. *Pli: The Warwick Journal of Philosophy*, 10: 174–90.

Badiou, Alain (2001a). *Ethics: An Essay on the Understanding of Evil*. London: Verso.

Badiou, Alain (2001b). 'The Ethic of Truths: Construction and Potency'. *Pli: The Warwick Journal of Philosophy*, 12: 247–55.

Badiou, Alain (2001c). 'Who is Nietzsche?' *Pli: The Warwick Journal of Philosophy*, 11: 1–11.

Badiou, Alain (2002). 'Existence and Death'. *Discourse*, 24(1): 63–73.
Badiou, Alain (2003a). *Saint Paul: The Foundation of Universalism*. Stanford, CA: Stanford University Press.
Badiou, Alain (2003b). 'Beyond Formalisation' [interview with Peter Hallward]. *Angelaki*, 8(2): 111–36.
Badiou, Alain (2004). *Infinite Thought: Truth and the Return to Philosophy*. London: Continuum.
Badiou, Alain (2005a). 'Universal Truths and the Question of Religion' [interview with Adam S. Miller]. *Journal of Philosophy and Scripture*, 3 (1): 38–42.
Badiou, Alain (2005b). 'The Adventure of French Philosophy'. *New Left Review*, 35: 67–77.
Badiou, Alain (2005c). *Metapolitics*. London: Verso.
Badiou, Alain (2006a). *Being and Event*. London: Continuum.
Badiou, Alain (2006b). *Theoretical Writings*. London: Continuum.
Badiou, Alain (2006c). *Briefings on Existence. A Short Treatise on Transitory Ontology*. Albany, NY: SUNY Press.
Badiou, Alain (2006d). 'Plato, our Dear Plato!' *Angelaki*, 11(3): 39–41.
Badiou, Alain (2007a). *The Concept of Model*. Melbourne: re.press.
Badiou, Alain (2007b). 'Philosophy, Sciences, Mathematics: Interview with Alain Badiou'. *Collapse*, I: 11–26.
Badiou, Alain (2007c). 'New Horizons in Mathematics as a Philosophical Condition: An Interview with Alain Badiou'. *Parrhesia*, 3: 1–11.
Badiou, Alain (2008a). *Number and Numbers*. Malden, MA: Polity.
Badiou, Alain (2008b). *Conditions*. London: Continuum.
Badiou, Alain (2009). *Logics of Worlds*. London: Continuum.
Badiou, Alain (2012). 'The Concept of Change: Mathematics and Vitalism'. Q&A session at the European Graduate School. Online video, www.youtube.com/watch?v=Anxf1u6ikAk (accessed 1 May 2014).
Baki, Burhanuddin (2014). *Badiou's Being and Event and the Mathematics of Set Theory*. London: Bloomsbury.
Balibar, Etienne (1978). 'From Bachelard to Althusser: The Concept of Epistemological Break'. *Economy and Society*, 7(3): 207–37.
Bangu, Sorin (2012). *The Applicability of Mathematics in Science: Indispensability and Ontology*. Basingstoke: Palgrave Macmillan.
Bartlett, A.J. and Clemens, Justin (eds) (2010). *Alain Badiou: Key Concepts*. Durham: Acumen.
Beaney, Michael (eds) (1997). *The Frege Reader*. Oxford: Blackwell.
Beiser, Frederick (1987). *The Fate of Reason: German Philosophy from Kant to Fichte*. Cambridge, MA: Harvard University Press.
Beiser, Frederick (2011). *The German Historicist Tradition*. Oxford: Oxford University Press.
Benacerraf, Paul (1965). 'What Numbers Could Not Be'. *Philosophical Review*, 74(1): 47–73.
Benacerraf, Paul (1973). 'Mathematical Truth'. *The Journal of Philosophy*, 70(19): 661–79.
Benacerraf, Paul and Putnam, Hilary (eds) (1983). *Philosophy of Mathematics: Selected Readings*. Oxford: Oxford University Press.
Berlin, Isaiah (1980). *Against the Current: Essays in the History of Ideas*. New York: Viking Press.

Bhaskar, Roy (2008) [1975]. *A Realist Theory of Science*. London and New York: Routledge.
Bird, Alexander (2007). 'The Regress of Pure Powers?'. *The Philosophical Quarterly*, 57 (229): 513–34.
Bitbol, Michel and Kerszberg, Pierre and Petitot, Jean (eds) (2009). *Constituting Objectivity: Transcendental Perspectives on Modern Science*. Dordrecht: Springer.
Black, Max (1952). 'The Identity of Indiscernibles'. *Mind*, 61: 153–6.
Black, Max (1962). *Models and Metaphors: Studies in Language and Philosophy*. Ithaca, NY: Cornell University Press.
Bloor, David (1976). *Knowledge and Social Imagery*. London: Routledge & Kegan Paul.
Boden, Margaret (2004) [1991]. *The Creative Mind: Myths and Mechanisms*. London and New York: Routledge.
Boghossian, Paul and Peacocke, Christopher (2000). *New Essays on the A Priori*. Oxford: Clarendon Press.
Bokulich, Alisa and Bokulich, Peter (eds) (2011). *Scientific Structuralism*. Dordrecht: Springer.
Boniolo, Giovanni, Budinich, Paolo and Trobok, Majda (eds) (2005). *The Role of Mathematics in Physical Sciences: Interdisciplinary and Philosophical Aspects*. Dordrecht: Springer.
Bosteels, Bruno (2008). 'Radical Antiphilosophy'. *Filozofski vestnik*, XXIX(2): 155–87.
Boucher, David and Vincent, Andrew (2000). *British Idealism and Political Theory*. Edinburgh: Edinburgh University Press.
Bourbaki, Nicholas (1950). 'The Architecture of Mathematics'. *The American Mathematical Monthly*, 57(4): 221–32.
Bowden, Sean and Duffy, Simon (eds) (2012). *Badiou and Philosophy*. London and New York: Routledge.
Boyd, Richard (1973). 'Realism, Underdetermination, and a Causal Theory of Evidence'. *Noûs*, 7(1): 1–12.
Boyd, Richard (1980). 'Scientific Realism and Naturalistic Epistemology'. *PSA: Proceedings of the Biennial Meeting of the Philosophy of Science Association*, 2: 613–62.
Boyd, Richard (1983). 'On the Current Status of the Issue of Scientific Realism'. *Erkenntnis*, 19(1): 45–90.
Brading, Katherine and Castellani, Elena (eds) (2003). *Symmmetries in Physics: Philosophical Reflections*. Cambridge: Cambridge University Press.
Brandom, Robert (1994). *Making it Explicit. Reasoning, Representing, and Discursive Commitment*. Cambridge, MA: Harvard University Press.
Brandom, Robert (2002). *Tales of the Mighty Dead: Historical Essays in the Metaphysics of Intentionality*. Cambridge, MA: Harvard University Press.
Brassier, Ray (2005). 'Badiou's Materialist Epistemology of Mathematics'. *Angelaki*, 10(2): 135–50.
Brassier, Ray (2007). *Nihil Unbound: Enlightenment and Extinction*. Basingstoke: Palgrave.
Brassier, Ray (2011). 'The View From Nowhere'. *Identities: Journal for Politics, Gender and Culture*, 8(2): 7–23.
Brassier, Ray (2012). 'How to Train an Animal that Makes Inferences: Sellars on Rules and Regularities'. Paper presented at *The Human Animal in Politics, Science,*

and Psychoanalysis, KW Institute for Contemporary Art, Berlin, 16–17 December. Online video, http://vimeo.com/35371780 (accessed 1 May 2014).
Braver, Lee (2007). *A Thing of This World: A History of Continental Anti-Realism*. Evanston, IL: Northwestern University Press.
Braver, Lee (2012). 'A Brief History of Continental Realism'. *Continental Philosophy Review*, 45(2): 261–89.
Braver, Lee (2013). 'On Not Settling the Issue of Realism'. *Speculations*, IV: 9–14.
Brown, James Robert (2008). *Philosophy of Mathematics: A Contemporary Introduction to the World of Proofs and Pictures*. London and New York: Routledge.
Brown, James Robert (2012). *Platonism, Naturalism and Mathematical Knowledge*. London and New York: Routledge.
Brown, Richard E. and Milner Peter M. (2003). 'The Legacy of Donald O. Hebb: More than the Hebb Synapse'. *Nature Reviews. Neuroscience*, 4(12): 1013–19.
Bryant, Levi and Srnicek, Nick and Harman, Graham (eds) (2011). *The Speculative Turn: Continental Realism and Materialism*. Melbourne: re.press.
Bueno, Otávio (1999). 'What Is Structural Empiricism? Scientific Change in an Empiricist Setting'. *Erkenntnis*, 50: 59–85.
Bueno, Otávio (2000). 'Empiricism, Scientific Change and Mathematical Change'. *Studies in History and Philosophy of Science*, 31(2): 269–96.
Bueno, Otávio and French, Steven and Ladyman, James (2002). 'On Representing the Relationship between the Mathematical and the Empirical'. *Philosophy of Science*, 69: 497–518.
Bueno, Otávio and Linnebo, Øysten (eds) (2009). *New Waves in the Philosophy of Mathematics*. Basingstoke: Palgrave.
Burgess, John and Rosen, Gideon (1997). *A Subject with No Object: Strategies for Nominalistic Interpretation of Mathematics*. Oxford: Clarendon Press.
Busch, Jacob (2003). 'What Structures Could Not Be'. *International Studies in the Philosophy of Science*, 17 (3): 211–25.
Butterfield, Jeremy and Pagonis, Constantine (1999). *From Physics to Philosophy*. Cambridge: Cambridge University Press.
Calhoun, Craig and Juergensmeyer, Mark and VanAntwerpen, Jonathan (2011). *Rethinking Secularism*. Oxford: Oxford University Press.
Calude, Cristian S. and Chaitin, Gregory (1999). 'Randomness Everywhere'. *Nature*, 400(July): 319–20.
Canguilhem, Georges (1978). *On the Normal and the Pathological*. Dordrecht, Boston and New York: Reidel.
Cantor, Georg (1955). *Contributions to the Founding of the Theory of Transfinite Numbers*. Mineola, NY: Dover Publications.
Cao, Tian Yu (2003a). 'Structural Realism and the Interpretation of Quantum Field Theory'. *Synthese*, 136(1): 3–24.
Cao, Tian Yu (2003b). 'Can We Dissolve Physical Entities into Mathematical Structures?' *Synthese*, 136(1): 57–71.
Cao, Tian Yu (2010). *From Current Algebra to Quantum Chromodynamics: A Case For Structural Realism*. Cambridge: Cambridge University Press.
Caputo, John and Vattimo, Gianni (2007). *After the Death of God*. Edited by Jeffrey W. Robbins. New York: Columbia University Press.
Carey, Susan (2009). *The Origin of Concepts*. Oxford: Oxford University Press.
Carnap, Rudolf (1939). *Foundations of Logic and Mathematics*. Chicago: The University of Chicago Press.

Carnap, Rudolf (1950). 'Empiricism, Semantics, and Ontology'. *Revue Internationale de Philosophie*, 4: 20–40.

Carroll, John W. and Markosian, Ned (2010). *An Introduction to Metaphysics*. Cambridge: Cambridge University Press.

Cartwright, Nancy (1983). *How the Laws of Physics Lie*. Oxford: Oxford University Press.

Cartwright, Nancy (1999). *The Dappled World: A Study of the Boundaries of Science*. Cambridge: Cambridge University Press.

Castellani, Elena (ed.) (1998). *Interpreting Bodies: Classical and Quantum Objects in Modern Physics*. Princeton, NJ: Princeton University Press.

Cazeaux, Clive (2007). *Metaphor and Continental Philosophy: From Kant to Derrida*. London and New York: Routledge.

Chaitin, Gregory (1975). 'Randomness and Mathematical Proof'. *Scientific American*, 232(5): 47–52.

Chaitin, Gregory (1987). *Algorithmic Information Theory*. Cambridge and New York: Cambridge University Press.

Chaitin, Gregory (1998). *The Limits of Mathematics*. Dordrecht: Springer.

Chaitin, Gregory (1999). *The Unknowable*. Dordrecht: Springer.

Chaitin, Gregory (2007). *MetaMaths: The Quest for Omega*. London: Atlantic Books.

Chaitin, Gregory (2012). *Proving Darwin: Making Biology Mathematical*. New York: Vintage.

Chakravartty, Anjan (2007). *A Metaphysics for Scientific Realism: Knowing the Unobservable*. Cambridge: Cambridge University Press.

Chalmers, David and Manley, David and Wasserman, Ryan (eds) (2009). *Metametaphysics: New Essays on the Foundations of Ontology*. Oxford: Oxford University Press.

Changeux, Jean-Pierre and Connes, Alain (1995). *Conversations on Mathematics, Matter, and Mind*. Princeton, NJ: Princeton University Press.

Chihara, Charles (1990). *Constructibility and Mathematical Existence*. Oxford: Oxford University Press.

Chomsky, Noam (2007). *On Language*. New York: The New Press.

Churchland, Patricia (1987). 'Epistemology in the Age of Neuroscience'. *The Journal of Philosophy*, 84(10): 544–53.

Churchland, Paul and Hooker, Clifford Alan (1985). *Images of Science*. Chicago: The University of Chicago Press.

Clark, Andy (2008). *Supersizing the Mind: Embodiment, Action, and Cognitive Extension*. Oxford: Oxford University Press.

Clayton, Philip and Simpson, Zachary (eds) (2006). *The Oxford Handbook of Religion and Science*. Oxford: Oxford University Press.

Clemens, Justin (2005). 'Doubles of Nothing: The Problem of Binding Truth to Being in the World of Alain Badiou'. *Filozofski Vestnik*, XXVI(2): 97–111.

Coffa, Alberto J. (1991). *The Semantic Tradition from Kant to Carnap: To the Vienna Station*. Cambridge: Cambridge University Press.

Cohen, Floris (1994). *The Scientific Revolution: A Historiographical Inquiry*. Chicago: The University of Chicago Press.

Cohen, Floris (2010). *How Modern Science Came into the World: Four Civilizations, One 17th-Century Breakthrough*. Amsterdam: Amsterdam University Press.

Cohen, Paul J. (1966). *Set Theory and the Continuum Hypothesis*. New York and Amsterdam: W. A. Benjamin, Inc.
Collingwood, Robin George (1940). *An Essay on Metaphysics*. Oxford: Clarendon Press.
Colyvan, Mark (2001a). *The Indispensability of Mathematics*. Oxford: Oxford University Press.
Colyvan, Mark (2001b). 'The Miracle of Applied Mathematics'. *Synthese*, 127: 265–77.
Constant, E.W. II (1980). *The Origins of the Turbojet Revolution*. Baltimore: Johns Hopkins University Press.
Corcoran, Steven (ed.) (forthcoming). *The Badiou Dictionary*. Edinburgh: Edinburgh University Press.
Cornelli, Gabriele (2013). *In Search of Pythagoreanism: Pythagoreanism as an Historiographical Category*. Berlin and New York: de Gruyter.
Corry, Leo (1992). 'Nicolas Bourbaki and the Concept of Mathematical Structure'. *Synthese*, 92(3): 315–48.
Curd, Martin and Cover, J.A. (eds) (1998). *Philosophy of Science: The Central Issues*. London and New York: Norton & Company.
D'Oro, Giuseppina (2002). *Collingwood and the Metaphysics of Experience*. London and New York: Routledge.
Dahlbom, Bo (1993). *Dennett and His Critics: Demystifying Mind*. Malden, MA: Blackwell.
Daston, Lorraine and Vidal, Fernando (eds) (2004). *The Moral Authority of Nature*. Chicago: The University of Chicago Press.
Dauben, Joseph Warren (1990). *Georg Cantor: His Mathematics and Philosophy of the Infinite*. Princeton, NJ: Princeton University Press.
De Caro, Mario and Macarthur, David (eds) (2004). *Naturalism in Question*. Cambridge, MA: Harvard University Press.
De Caro, Mario and Macarthur, David (eds) (2010). *Naturalism and Normativity*. New York: Columbia University Press.
Dehaene, Stanislas (2011) [1997]. *The Number Sense: How the Mind Creates Mathematics*. Revised and Expanded Edition. Oxford: Oxford University Press.
Dehaene, Stanislas and Brannon, Elizabeth (eds) (2011). *Space, Time and Number in the Brain: Searching for the Foundations of Mathematical Thought*. London: Academic Press.
Dehaene, Stanislas and Changeux, Jean Pierre (1993). 'Development of Elementary Numerical Abilities: A Neuronal Model'. *Journal of Cognitive Neuroscience*, 5(4): 390–407.
Deleuze, Gilles (1991). *Empiricism and Subjectivity: An Essay on Hume's Theory of Human Nature*. New York: Columbia University Press.
Deleuze, Gilles (2001). *Difference and Repetition*. London: Continuum.
Deleuze, Gilles and Guattari, Félix (1994). *What is Philosophy?* New York: Columbia University Press.
Della Rocca, Michael (2005). 'Two Spheres, Twenty Spheres, and the Identity of Indiscernibles'. *Pacific Philosophical Quarterly*, 86: 480–92.
Della Rocca, Michael (2008). *Spinoza*. London and New York: Routledge.
Dennett, Daniel (1989). *The Intentional Stance*. Cambridge, MA: MIT Press.
Dennett, Daniel (1991). 'Real Patterns'. *The Journal of Philosophy*, 88(1): 27–51.

Dennett, Daniel (1996). *Darwin's Dangerous Idea: Evolution and the Meaning of Life*. London: Penguin.

Depoortere, Frederiek (2009). *Badiou and Theology*. London: Continuum.

Derrida, Jacques (1974). *Of Grammatology*. Baltimore and London: Johns Hopkins University Press.

Derrida, Jacques (1981). *Positions*. Chicago: The University of Chicago Press.

Derrida, Jacques (1982). *Margins of Philosophy*. Brighton: Harvester Press.

Derrida, Jacques (1989). *Edmund Husserl's Origin of Geometry: An Introduction*. Lincoln, NE and London: University of Nebraska Press.

Derrida, Jacques (1995). 'A Certain "Madness" Must Watch Over Thinking', Interview with François Ewald. *Educational Theory*, 45(3): 273–91.

Derrida, Jacques (2004). *Dissemination*. London: Continuum.

Derrida, Jacques (2007). *Psyche, Inventions of the Other, Volume I*. Stanford, CA: Stanford University Press.

Descartes, René (1996). *Meditations on First Philosophy*. Cambridge: Cambridge University Press.

Descartes, René (2000). *Philosophical Essays and Correspondence*. Indianapolis, IN and Cambridge, MA: Hackett.

Deutsch, David (2012). 'Creative Blocks. The Very Laws of Physics Imply that Artificial Intelligence Must be Possible. What's Holding Us Up?'. *Aeon Magazine*. Online http://aeon.co/magazine/technology/david-deutsch-artificial-intelligence/ (accessed 1 May 2014).

Devitt, Michael (1997) [1984]. *Realism and Truth*. Princeton, NJ: Princeton University Press.

Dipert, Randall R. (1997). 'The Mathematical Structure of the World: The World as Graph'. *The Journal of Philosophy*, 94(7): 329–58.

Domski, Mary and Dickson, Michael (eds) (2000). *Discourse on a New Method*. Chicago and La Salle, IL: Open Court.

Dorato, Mauro (2000). 'Substantivalism, Relationism, and Structural Spacetime Realism'. *Foundations of Physics*, 30 (10): 1605–28.

Duhem, Pierre (1991) [1914]. *The Aim and Structure of Physical Theory*. Princeton, NJ: Princeton University Press.

Dummett, Michael (1978). *Truth and Other Enigmas*. Cambridge, MA: Harvard University Press.

Dummett, Michael (1991). *Frege: Philosophy of Mathematics*. London: Duckworth.

Echeverria, Javier and Ibarra, Andoni and Mormann, Thomas (eds) (1992). *The Space of Mathematics*. Berlin and New York: de Gruyter.

Eddington, Arthur (1929). *The Nature of the Physical World*. New York: Macmillan.

Ernest, Paul (1998). *Social Constructivism as a Philosophy of Mathematics*. New York: SUNY Press.

Esfeld, Michael and Lam, Vincent (2008). 'Moderate Structural Realism About Space-time'. *Synthese*, 160(1): 27–46.

Ewald, William Bragg (1996). *From Kant to Hilbert. A Source Book in the Foundations of Mathematics. Volume 2*. Oxford: Clarendon Press.

Eyers, Tom (2013). *Post-Rationalism: Psychoanalysis, Epistemology, and Marxism in Post-War France*. London: Bloomsbury.

Feigl, Herbert and Brodbeck, May (eds) (1953). *Readings in the Philosophy of Science*. New York: Appleton Century Crofts.

Feigl, Herbert and Maxwell, Grover (eds) (1962). *Scientific Explanation, Space and Time. Minnesota Studies in the Philosophy of Science, Vol. 3.* Minneapolis, MN: University of Minnesota Press.

Feltham, Oliver (2008). *Alain Badiou: Live Theory.* London: Continuum.

Field, Hartry (1980). *Science without Numbers: A Defence of Nominalism.* Princeton, NJ: Princeton University Press.

Floridi, Luciano (2008). 'A Defence of Informational Structural Realism'. *Synthese,* 161(2): 219–53.

Foucault, Michel (1980). *Power/Knowledge: Selected Interviews and Other Writings, 1972–1977.* New York: Pantheon Books.

Foucault, Michel (2002) [1966]. *The Order of Things: An Archaeology of the Human Sciences.* London and New York: Routledge.

Frege, Gottlob (1980). *The Philosophical and Mathematical Correspondence.* Oxford: Basil Blackwell.

French, Steven (2000). 'The Reasonable Effectiveness of Mathematics: Partial Structures and the Application of Group Theory to Physics'. *Synthese,* 125(1): 103–20.

French, Steven (2003). 'Scribbling on the Blank Sheet: Eddington's Structuralist Conception of Objects'. *Studies in History and Philosophy of Science Part B: Studies in History and Philosophy of Modern Physics,* 34(2): 227–59.

French, Steven (2006). 'Structure as a Weapon of the Realist'. *Proceedings of the Aristotelian Society,* 106(1): 170–87.

French, Steven (2010). 'The Interdependence of Structure, Objects and Dependence'. *Synthese,* 175 (S1): 89–109.French, Steven (2011). 'Shifting to Structures in Physics and Biology: A Prophylactic for Promiscuous Realism'. *Studies in History and Philosophy of Biological and Biomedical Sciences,* 42(2): 164–73.

French, Steven (2014). *The Structure of the World. Metaphysics and Representation.* Oxford: Oxford University Press.

French, Steven and Krause, Decio (2006). *Identity in Physics: A Historical, Philosophical, and Formal Analysis.* Oxford: Oxford University Press.

French, Steven, and Ladyman, James (2003a). 'Remodelling Structural Realism: Quantum Physics and the Metaphysics of Structure'. *Synthese,* 136(1): 31–56.

French, Steven, and Ladyman, James (2003b). 'The Dissolution of Objects: Between Platonism and Phenomenalism'. *Synthese,* 136(1): 73–7.

French, Steven and McKenzie, Kerry (2012). 'Thinking Outside the (Tool)Box: Towards a More Productive Engagement Between Metaphysics and Philosophy of Physics'. *European Journal of Analytic Philosophy,* 8(1): 42–59.

French, Steven and Redhead, Michael (1988). 'Quantum Physics and the Identity of Indiscernibles'. *The British Journal for the Philosophy of Science,* 39(2): 233–46.

Friedman, Michael (1999). *Reconsidering Logical Positivism.* Cambridge: Cambridge University Press.

Friedman, Michael (2001). *Dynamics of Reason.* Chicago: The University of Chicago Press.

Friedman, Michael (2009). 'Carnap on Theoretical Terms: Structuralism Without Metaphysics'. *Synthese,* 180(2): 249–63.

Frigg, Roman, and Votsis, Ioannis (2011). 'Everything You Always Wanted to Know about Structural Realism but were Afraid to Ask'. *European Journal for Philosophy of Science,* 1(2): 227–76.

Gabriel, Markus (2011). *Transcendental Ontology*. London and New York: Routledge.

Galilei, Galileo (1954). *Dialogue Concerning Two New Sciences*. New York: Dover.

Garfield, Jay L. (2012). 'Sellarsian Synopsis: Integrating the Images'. *Humana. Mente: Journal of Philosophical Studies*, 21: 103–21.

Gaukroger, Stephen (2006). *The Emergence of a Scientific Culture. Science and the Shaping of Modernity, 1210–1685*. Oxford: Clarendon Press.

Geroulanos, Stephanos (2010). *An Atheism that is not Humanist Emerges in French Thought*. Stanford, CA: Stanford University Press.

Gibson, Roger F. (ed.) (2004). *The Cambridge Companion to Quine*. Cambridge: Cambridge University Press.

Giere, Ronald N. (ed.) (1992). *Cognitive Models of Science*. Minneapolis, MN: University of Minnesota Press.

Gillespie, Sam (2008). *The Mathematics of Novelty: Badiou's Minimalist Metaphysics*. Melbourne: re.press.

Ginev, Dimitri (2011). *The Tenets of Cognitive Existentialism*. Athens, OH: Ohio University Press.

Giniammi, Michele (2011). 'The Unreasonable Effectiveness. The Philosophical Problem of Applicability of Mathematics'. *Rivista Italiana Di Filosofia Analitica*, 2(2): 26–40.

Gödel, Kurt (1990). *Collected Works, Volume II*: Oxford: Oxford University Press.

Goldman, Alvin (1967). 'A Causal Theory of Knowing', *The Journal of Philosophy*, 64 (12): 357–72.

Goodenough, Ursula (1998). The Sacred Depths of Nature. Oxford: Oxford University Press.

Goodman, Nelson (1978). *Ways of Worldmaking*. Cambridge, MA: Hackett.

Gourgouris, Stathis (2013). *Lessons in Secular Criticism*. New York: Fordham University Press.

Gower, Barry (2000). 'Cassirer, Schlick and "Structural" Realism: The Philosophy of the Exact Sciences in the Background to Early Logical Empiricism'. *British Journal for the History of Philosophy*, 8(1): 71–106.

Gowers, Timothy and Barrow-Green, June and Leader, Imre (eds) (2008). *The Princeton Companion of Mathematics*. Princeton, NJ: Princeton University Press.

Grabiner, Judith V. (1981). *The Origins of Cauchy's Rigorous Calculus*. Cambridge, MA: MIT Press.

Grice, H.P. (1961). 'The Causal Theory of Perception'. *Proceedings of the Aristotelian Society*, xxxv: 121–53.

Gutting, Gary (1987). 'Gaston Bachelard's Philosophy of Science', *International Studies in the Philosophy of Science*, 2(1): 55–71.

Gutting, Gary (2011). *Thinking the Impossible: French Philosophy since 1960*. Oxford: Oxford University Press.

Hacking, Ian (1981). 'Do We See Through a Microscope?', *Pacific Philosophical Quarterly*, 62(4): 305–22.

Hacking, Ian (1999). *The Social Construction of What?* Cambridge, MA: Harvard University Press.

Hacking, Ian (2000). 'What Mathematics Has Done to Some and Only Some Philosophers'. *Proceedings of the British Academy*, 103: 83–138.

Hallett, Michael (1984). *Cantorian Set Theory and Limitation of Size*. Oxford: Oxford University Press.

Hallward, Peter (2003). *Badiou: A Subject to Truth*. Minneapolis, MN and London: University of Minnesota Press.

Hallward, Peter (ed.) (2004). *Think Again: Alain Badiou and the Future of Philosophy*. London: Continuum.

Hallward, Peter (2005). 'Depending on Inconsistency: Badiou's Answer to the "Guiding Question of All Contemporary Philosophy"'. *Polygraph*, 17: 7–21.

Hallward, Peter (2006). *Out of This World: Deleuze and the Philosophy of Creation*. London and New York: Verso.

Hallward, Peter (2008). 'Order and Event: On Badiou's *Logics of Worlds*'. *New Left Review*, 53: 97–122.

Hallward, Peter and Peden, Knox (eds) (2012). *Concept and Form*, Volume 1: Selections from *Cahiers pour l'Analyse*. London: Verso.

Halperin, David M. (1998). 'Forgetting Foucault: Acts, Identities, and the History of Sexuality'. *Representations*, 63: 93–120.

Hanson, Norwood Russell (1958). *Patterns of Discovery: An Inquiry into the Conceptual Foundations of Science*. Cambridge: Cambridge University Press.

Hardy, G.H. (2005) [1940]. *A Mathematician's Apology*. University of Alberta Mathematical Sciences Society. Online, www.math.ualberta.ca/~mss/misc/A Mathematician's Apology.pdf (accessed 1 May 2014).

Harman, Gilbert (1965). 'The Inference to the Best Explanation'. *The Philosophical Review*, 74(1): 88–95.

Hebb, Donald O. (2002) [1949]. *The Organization of Behavior: A Neuropsychological Theory*. Mahwah, NJ and London: Lawrence Erlbaum Associates.

Hegel, G.W.F. (1990). *Lectures on the History of Philosophy, Volume III: Medieval and Modern Philosophy*. Berkeley, CA: University of California Press.

Hegel, G.W.F. (2010). *The Science of Logic*. Cambridge: Cambridge University Press.

Heidegger, Martin (1968). *What is Called Thinking?* New York: Harper & Row.

Heidegger, Martin (1982). *The Basic Problems of Phenomenology*. Bloomington and Indianapolis, IN: Indiana University Press.

Heidegger, Martin (1993). *Basic Writings*. San Francisco: Harper.

Heidegger, Martin (2003). *Philosophical and Political Writings*. London: Continuum.

Hellman, Geoffrey (1989) *Mathematics Without Numbers*. Oxford: Oxford University Press.

Hellyer, Marcus (ed.) (2003). *The Scientific Revolution: The Essential Readings*. Malden, MA: Blackwell.

Hesse, Mary (1988). 'The Cognitive Claims of Metaphor'. *The Journal of Speculative Philosophy*, 2(1): 1–16.

Hey, Anthony J.G. (ed.) (1999). *Feynman and Computation: Exploring the Limits of Computers*. Reading, MA: Perseus Books.

Hilbert, David and Ackermann, Wilhelm (1950) [1938]. *Principles of Mathematical Logic*. New York: Chelsea Publishing Company.

Hilgevoord Jan (ed.) (1994). *Physics and Our View of the World*. Cambridge: Cambridge University Press.

Hitchcock, Christopher (ed.) (2004). *Contemporary Debates in Philosophy of Science*. Oxford: Blackwell.

Horwich, Paul (2004). *From a Deflationary Point of View*. Oxford: Oxford University Press.

Hut, Piet and Alford, Mark and Tegmark, Max (2006). 'On Math, Matter and Mind'. *Foundations of Physics*, 36(6): 765–94.

Ippoliti, Emiliano (2008). *Inferenze Ampliative: Visualizzazione, Analogia e Rappresentazioni Multiple*. Morrisville, NC: Lulu Press.

Israel, Jonathan (2001). *Radical Enlightenment. Philosophy and the Making of Modernity 1650–1750*. Oxford: Oxford University Press.

Israel, Jonathan (2006). *Enlightenment Contested Philosophy Modernity and the Emancipation of Man 1670–1752*. Oxford: Oxford University Press.

Israel, Jonathan (2011). *Democratic Enlightenment. Philosophy, Revolution and Human Rights 1750–1790*. Oxford: Oxford University Press.

Jacquette, Dale (2002). *Ontology*. Durham: Acumen.

James, William (1955) [1907]. *Pragmatism, and Four Essays from The Meaning of Truth*. New York: Meridian Books.

Janicaud, Dominique (ed.) (2000). *Phenomenology and the 'Theological Turn': The French Debate*. New York: Fordham University Press.

Johnston, Adrian (2008a). 'What Matter(s) in Ontology. Alain Badiou, the Hebb-Event, and Materialism Split from Within'. *Angelaki*, 13(1): 27–49.

Johnston, Adrian (2008b). 'Phantom of Consistency: Alain Badiou and Kantian Transcendental Idealism'. *Continental Philosophy Review*, 41(3): 345–66.

Johnston, Adrian (2011). 'Materialism, Subjectivity and the Outcome of French Philosophy'. Interview with Michael O Neill Burns and Brian Anthony Smith. *Cosmos and History: The Journal of Natural and Social Philosophy*, 7(1): 167–81.

Johnston, Adrian (2013). 'Points of Forced Freedom: Eleven (More) Theses on Materialism'. *Speculations IV*, 91–8.

Josephson, John and Josephson, Susan (1994). *Abductive Inference: Computation, Philosophy and Technology*. Cambridge: Cambridge University Press.

Kant, Immanuel (2004). *Prolegomena to Any Future Metaphysics*. Cambridge: Cambridge University Press.

Kantorovich, Aharon (2009). 'Ontic Structuralism and the Symmetries of Particle Physics'. *Journal for General Philosophy of Science*, 40(1): 73–84.

Katz, Jerrold J. (1998). *Realistic Rationalism*. Cambridge, MA: MIT Press.

Kincaid, Harold (2008). 'Structural Realism and the Social Sciences'. *Philosophy of Science*, 75 (5): 720–31.

Kitcher, Philip (1983). *The Nature of Mathematical Knowledge*. Oxford: Oxford University Press.

Kitcher, Philip (2012). *Preludes to Pragmatism: Toward a Reconstruction of Philosophy*. Oxford: Oxford University Press.

Kitcher, Philip, and Salmon, Wesley (1987). 'van Fraassen on Explanation'. *The Journal of Philosophy*, 84(6): 315–30.

Klee, Robert (ed.) (1998). *Scientific Inquiry: Readings in the Philosophy of Science*. Oxford: Oxford University Press.

Klein, Peter (1998). 'Foundationalism and the Infinite Regress of Reasons'. *Philosophy and Phenomenological Research*, 58(4): 919–25.

Klein, Peter (2006). 'Human Knowledge and the Infinite Progress of Reasoning'. *Philosophical Studies*, 134(1): 1–17.

Kline, Morris (1980). *Mathematics: The Loss of Certainty*. Oxford: Oxford University Press.

Kosso, Peter (2011). *A Summary of Scientific Method*. Dordrecht: Springer.

Koyré, Alexandre (1957). *From the Closed World to the Infinite Universe*. Baltimore, MD: The Johns Hopkins Press.

Kuhlmann, Meinard (2010). *The Ultimate Constituents of the Material World: In Search for an Ontology for Fundamental Physics*. Frankfurt: Ontos Verlag.

Kuhlmann, Meinard and Lyre, Holger and Wayne, Andrew (eds) (2002). *Ontological Aspects of Quantum Field Theory*. London and Singapore: World Scientific.

Kuhn, Thomas (2012) [1962]. *The Structure of Scientific Revolutions*. Fourth Edition. Chicago: The University of Chicago Press.

Kukla, André (1998). *Studies in Scientific Realism*. Oxford: Oxford University Press.

Lacan, Jacques (2002). *Écrits*. New York and London: W.W. Norton & Company.

Ladyman, James (1998). 'What is Structural Realism?' *Studies in History and Philosophy of Science*, 29: 409–24.

Ladyman, James. (2000). 'What's Really Wrong with Constructive Empiricism? Van Fraassen and the Metaphysics of Modality'. *British Journal for the Philosophy of Science*, 51: 837–56.

Ladyman, James. (2001). 'Science, Metaphysics and Structural Realism'. *Philosophica*, 67(1): 57–76.

Ladyman, James (2005). 'Mathematical Structuralism and the Identity of Indiscernibles'. *Analysis*, 65(287): 218–21.

Ladyman, James (2007). 'On the Identity and Diversity of Objects in a Structure'. *Proceedings of the Aristotelian Society Supplementary Volume LXXXI* (1911): 23–43.

Ladyman, James (2008). 'Structural Realism and the Relationship between the Special Sciences and Physics'. *Philosophy of Science*, 75: 744–55.

Ladyman, James (2009). 'Who's Afraid of Scientism?'. *Collapse*, V, 135–85.

Ladyman, James (2013). 'Structural Realism' *Stanford Encyclopedia of Philosophy*, summer. Edited by Edward N. Zalta. Online, http://plato.stanford.edu/archives/sum2013/entries/structural-realism (accessed 1 May 2014).

Ladyman, James, and Bigaj, Tomasz (2010). 'The Principle of the Identity of Indiscernibles and Quantum Mechanics'. *Philosophy of Science*, 77(1): 117–36.

Ladyman, James and Douven, Igor and Horsten, Leon and van Fraassen, Bas (1997). 'A Defence of van Fraassen's Critique of Abductive Inference: Reply to Psillos'. *The Philosophical Quarterly*, 47(188): 305–21.

Ladyman, James and Ross, Don (2007). *Every Thing Must Go: Metaphysics Naturalised*. Oxford: Oxford University Press.

Ladyman, James and Ross, Don and Kincaid, Harold (eds) (2013). *Scientific Metaphysics*. Oxford: Oxford University Press.

Lakatos, Imre (1976). *Proofs and Refutations. The Logic of Mathematical Discovery*. Cambridge: Cambridge University Press.

Lakoff, George and Johnson, Mark (1980). *Metaphors We Live By*. Chicago: The University of Chicago Press.

Lakoff, George and Núñez, Rafael E. (2000). *Where Mathematics Comes From: How the Embodied Mind Brings Mathematics into Being*. New York: Basic Books.

Landry Elaine M. and Rickles Dean P. (eds) (2012). *Structural Realism: Structure Object and Causality*. Dordrecht: Springer.

Langley, Pat and Simon, Herbert A. and Bradshaw, Gary L. and Zytkow, Jon M. (eds) (1987). *Scientific Discovery: Computational Explorations of the Creative Processes*. Cambridge, MA: MIT Press.

Langton, Rae (1998). *Kantian Humility: Our Ignorance of Things in Themselves.* Oxford: Clarendon Press.

Langton, Rae and Lewis, David (1998). 'Defining "Intrinsic"'. *Philosophy and Phenomenological Research*, LVIII (2): 333–45.

Lautman, Albert (2011). *Mathematics, Ideas and the Physical Real.* London: Continuum.

Lecourt, Dominic (1975). *Marxism and Epistemology: Bachelard, Canguilhem, and Foucault.* London: NLB.

Leibniz, Gottfried Wilhelm (1989). *Philosophical Papers and Letters.* Edited by Leroy E. Loemker. Dordrecht: Kluwer Academic Publishers.

Leplin, Jarret (1984). *Scientific Realism.* Los Angeles: University of California Press.

Leplin, Jarret (1997). *A Novel Defense of Scientific Realism.* Oxford: Oxford University Press.

Levinas, Emmanuel (1991). *Totality and Infinity.* Dordrecht: Kluwer Academic Publishers.

Levitin, Daniel J. (2002). *Foundations of Cognitive Psychology: Core Readings.* Cambridge, MA: MIT Press.

Lewis, David (1973). *Counterfactuals.* Oxford: Blackwell.

Lewis, David (1986). *On the Plurality of Worlds.* Oxford: Basil Blackwell.

Lindberg, David C. and Westman, Robert S. (eds) (1990). *Reappraisals of the Scientific Revolution.* Cambridge and New York: Cambridge University Press.

Linnebo, Øystein (2008). 'Structuralism and the Notion of Dependence'. *The Philosophical Quarterly*, 58(230): 59–79.

Lipton, Peter (2004) [1991]. *Inference to the Best Explanation.* London and New York: Routledge.

Livingston, Paul (2008). *Philosophy and the Vision of Language.* London and New York: Routledge.

Livingston, Paul (2011). *The Politics of Logic: Badiou, Wittgenstein, and the Consequences of Formalism.* London and New York: Routledge.

Livingston, Paul (2013). 'Realism and the Infinite'. *Speculations*, IV: 99–107.

Lloyd, Genevieve (2013). *Enlightenment Shadows.* Oxford: Oxford University Press.

Locke, John (1996). *An Essay Concerning Human Understanding.* Indianapolis, IN and Cambridge, MA: Hackett.

Lombrozo, Tania (2006). 'The Structure and Function of Explanations'. *Trends in Cognitive Sciences*, 10(10): 464–70.

Longo, Giuseppe and Baily, Francis (2011). *Mathematics and the Natural Sciences: The Physical Singularity of Life.* London: Imperial College Press.

Lovejoy, Arthur (2001) [1936]. *The Great Chain of Being.* Cambridge, MA and London: Harvard University Press.

Lucretius, Titus (2001). *On the Nature of Things.* Indianapolis, IN and Cambridge, MA: Hackett.

MacBride, Fraser (2006). *Identity and Modality.* Oxford. Oxford University Press.

McKenzie, Kerry (2012). *Physics Without Fundamentality: A Study of Anti-Fundamentalism about Particles and Laws in High-Energy Physics.* PhD thesis, University of Leeds.

Maddy, Penelope (1990). *Realism in Mathematics.* Oxford: Clarendon Press.

Maddy, Penelope (1997). *Naturalism in Mathematics.* Oxford: Clarendon Press.

Maddy, Penelope (2007). *Second Philosophy. A Naturalistic Method*. Oxford: Oxford University Press.
Maddy, Penelope (2011). *Defending the Axioms*. Oxford: Oxford University Press.
Magnani, Lorenzo (2001). *Abduction, Reason, and Science: Processes of Discovery and Explanation*. Dordrecht: Kluwer.
Magnani, Lorenzo (2009). *Abductive Cognition: The Epistemological and Eco-Cognitive Dimensions of Hypothetical Reasoning*. Dodrecht: Springer.
Magnani, Lorenzo and Li, Ping (eds) (2007). *Model-Based Reasoning in Science, Technology, and Medicine*. Dordrecht: Springer.
McArthur, Daniel (2002). 'Why Bachelard Is Not a Scientific Realist'. *The Philosophical Forum*, XXXIII (2): 159–72.
McDowell, John (1996). *Mind and World*. Cambridge, MA and London: Harvard University Press.
McMullin, Ernan (1992). *The Inference that Makes Science*. Milwaukee: Marquette University Press.
Meillassoux, Quentin (2008). *After Finitude*. London: Continuum.
Menary, Richard (eds) (2010). *The Extended Mind*. Cambridge, MA and London: MIT Press.
Mill, John Stuart (1981). *Autobiography and Literary Essays*. Toronto and Buffalo, NY: University of Toronto Press.
Moore, A.W. (1990). *The Infinite*. London and New York: Routledge.
Moore, A.W. (2012). *The Evolution of Modern Metaphysics: Making Sense of Things*. Cambridge: Cambridge University Press.
Morganti, Matteo (2004). 'On the Preferability of Epistemic Structural Realism'. *Synthese*, 142(1): 81–107.
Morrison, Margaret (2000). *Unifying Scientific Theories: Physical Concepts and Mathematical Structures*. Cambridge: Cambridge University Press.
Mumford, Stephen and Tugby, Matthew (eds) (2013). *Metaphysics of Science*. Oxford: Oxford University Press.
Murray, Michael (ed.) (1978). *Heidegger and Modern Philosophy: Critical Essays*. New Haven, CT and London: Yale University Press.
Nagel, Thomas (2012). *Mind and Cosmos: Why the Materialist Neo-Darwinian Conception of Nature Is Almost Certainly False*. Oxford: Oxford University Press.
Nersessian, Nancy (2008). *Creating Scientific Concepts*. Cambridge, MA: MIT Press.
Netz, Reviel (1999). *The Shaping of Deduction in Greek Mathematics: A Study in Cognitive History*. Cambridge: Cambridge University Press.
Netz, Reviel (2002). 'Counter Culture: Towards a History of Greek Numeracy'. *History of Science*, XL: 321–52.
Neurath, Maria and Cohen R.S. (eds) (1973). *Empiricism and Sociology*. Dordrecht: Reidel.
Nietzsche, Friedrich (2001). *The Gay Science*. Cambridge: Cambridge University Press.
Niiniluoto, Ilkka (1999). *Critical Scientific Realism*. Oxford: Oxford University Press.
Nola, Robert (ed.) (1988). *Realism and Relativism in Science*. Dordrecht: Kluwer Academic Publishing.
Nola, Robert and Sankey, Howard (2007). *Theories of Scientific Method*. Durham: Acumen.

Norris, Christopher (1997). *New Idols of the Cave: On the Limits of Anti-realism*. Manchester: Manchester University Press.

Norris, Christopher (2000). *Minding the Gap: Epistemology and Philosophy of Science in the Two Traditions*. Amherst, MA: University of Massachusetts Press.

Norris, Christopher (2004). *Philosophy of Language and the Challenge of Scientific Realism*. London and New York: Routledge.

Norris, Christopher (2009). *Badiou's Being and Event: A Reader's Guide*. London: Continuum.

Norris, Christopher (2010). *Re-thinking the Cogito: Naturalism, Rationalism and the Venture of Thought*. London: Continuum.

Norris, Christopher (2011). 'Tractatus Mathematicus-Philosophicus: On Alain Badiou's *Being and Event*'. *Speculations*, II: 7–48.

Norris, Christopher (2012). *Derrida, Badiou, and the Formal Imperative*. London: Continuum.

Nye, Mary Jo (ed.) (2002). *The Cambridge History of Science, Volume 5: The Modern Physical and Mathematical Sciences*. Cambridge: Cambridge University Press.

O'Shea, James (2007). *Wilfrid Sellars: Naturalism with a Normative Turn*. Cambridge and Malden, MA: Polity.

O'Shea, James (2011). 'How to Be a Kantian and a Naturalist about Human Knowledge: Sellars's Middle Way'' *Journal of Philosophical Research*, 36: 327–59.

Oderberg, D. S. (2010). 'The World Is Not an Asymmetric Graph' *Analysis*, 71(1): 3–10.

Oderberg, D. S. (2011). 'Graph Structuralism and Its Discontents: Rejoinder to Shackel'. *Analysis*, 72(1): 94–8.

Ortony, Anthony (ed.) (1979). *Metaphor and Thought*. Cambridge: Cambridge University Press.

Osborne, Peter (2007). 'Neo-Classic. Alain Badiou's Being and Event'. *Radical Philosophy*, 142: 19–29.

Osler, Margaret J. (ed.) (2000). *Rethinking the Scientific Revolution*. Cambridge: Cambridge University Press.

Panza, Marco and Sereni, Andrea (2013). *Plato's Problem: An Introduction to Mathematical Platonism*. Basingstoke: Palgrave Macmillan.

Parsons, Charles (1990). 'The Structuralist View of Mathematical Objects'. *Synthese*, 84(3): 303–46.

Parsons, Charles (2004). 'Structuralism and Metaphysics'. *The Philosophical Quarterly*, 54(214): 56–77.

Parsons, Charles (2008). *Mathematical Thought and Its Objects*. Cambridge: Cambridge University Press.

Pascal, Blaise (2004). *Pensees*. Indianapolis, IN: Hackett Publishing Company.

Paul, Elliot Samuel and Kauffman, Scott Barry (2014). *The Philosophy of Creativity: New Essays*. Oxford: Oxford University Press.

Peirce, Charles Sanders (1958–1966). *Collected Papers*. Vols 1–6 edited by Charles Hartshorne and Paul Weiss; vols 7–8 edited by A. W. Burks. Cambridge, MA: Belknap Press of Harvard University Press.

Penrose, Roger (2004). *The Road to Reality: A Complete Guide to the Laws of the Universe*. London: Jonathan Cape.

Phelps, Hollis (2013). *Alain Badiou: Between Theology and Anti-Theology*. Durham: Acumen.

Plantinga, Alvin (2011). *Where the Conflict Really Lies: Science, Religion, and Naturalism*, Oxford: Oxford University Press.

Pluth, Ed (2010). *Badiou: A Philosophy of the New*. Malden, MA: Polity.

Poincaré, Henri (1905) [1902]. *Science and Hypothesis*. London and Newcastle upon Tyne: Walter Scott Publishing.

Poincaré, Henri (1958) [1905]. *The Value of Science*. New York: Dover Publications.

Polkinghorne, John (1998). *Belief in God in an Age of Science*. New Haven, CT and London: Yale University Press.

Polkinghorne, John (ed.) (2011). *Meaning in Mathematics*. Oxford: Oxford University Press.

Popper, Karl (1962). *Conjectures and Refutations*. New York: Basic Books.

Porter, Theodore M. (1995). *Trust in Numbers: The Pursuit of Objectivity in Science and Public Life*. Princeton, NJ: Princeton University Press.

Post, H.R. (1971). 'Correspondence, Invariance and Heuristics'. *Studies in History and Philosophy of Science*, 2(3): 213–55.

Potter, Michael (2004). *Set Theory and its Philosophy: A Critical Introduction*. Oxford: Oxford University Press.

Prado, C. G. (ed.) (2003). *A House Divided: Comparing Analytic and Continental Philosophy*. Amherst, MA: Humanity Books.

Price, Huw and Corry, Richard (2007). *Causation, Physics, and the Constitution of Reality: Russell's Republic Revisited*. Oxford: Clarendon Press.

Pritchard, Paul (1995). *Plato's Philosophy of Mathematics*. Sankt Augustin: Academia Verlag.

Psillos, Stathis (1996). 'On van Fraassen's Critique of Abductive Reasoning'. *The Philosophical Quarterly*, 46(182): 31–47.

Psillos, Stathis (1997). 'How Not to Defend Constructive Empiricism: A Rejoinder'. *The Philosophical Quarterly*, 47(188): 369–72.

Psillos, Stathis (1999). *Scientific Realism: How Science Tracks Truth*. London and New York: Routledge.

Psillos, Stathis (2009). *Knowing the Structure of Nature: Essays on Realism and Explanation*. Basingstoke: Palgrave.

Psillos, Stathis and Curd, Martin (2008). *The Routledge Companion to Philosophy of Science*. London and New York: Routledge.

Purcell, Lynn Sebastian (2012). 'Two Paths to Infinite Thought: Alain Badiou and Jacques Derrida on the Question of the Whole'. *Cosmos and History: The Journal of Natural and Social Philosophy*, 8(1): 151–76.

Putnam, Hilary (1971). *Philosophy of Logic*. London: Allen & Unwin.

Putnam, Hilary (1975). *Mathematics, Matter and Method*. Cambridge: Cambridge University Press.

Putnam, Hilary (1978). *Meaning and the Moral Sciences*. London and New York: Routledge & Kegan Paul.

Putnam, Hilary (1981). *Reason, Truth and History*. Cambridge: Cambridge University Press.

Putnam, Hilary (2011) [1971]. *Philosophy of Logic*. London and New York: Routledge.

Quine, Willard Van Orman (1953). *From a Logical Point of View*. New York: Harper & Row.

Quine, Willard Van Orman (1981). *Theories and Things*. Cambridge, MA: Harvard University Press.

Quine, Willard Van Orman (1992). 'Structure and Nature', *The Journal of Philosophy*, 89(1): 5–9.

Rabinow, Paul (ed.) (1984). *The Foucault Reader*. New York: Pantheon Books.

Radner, Michael and Winokur, Stephen (1970). Analyses of Theories & Methods of Physics and Psychology. Minneapolis, MN: University of Minnesota Press.

Ratzinger, Joseph Aloisius (2009). 'Message of His Holiness Benedict XVI to Archibishop Rino Fisichella'. Online, www.vatican.va/holy_father/benedict_xvi/messages/pont-messages/2009/documents/hf_ben-xvi_mes_20091126_fisichella-telescopio_en.html (accessed 1 May 2013).

Reck, Erich H. (2003). 'Dedekind's Structuralism: An Interpretation and Partial Defense'. *Synthese*, 137: 369–419.

Reck, Erich H. and Price, Michael P. (2000). 'Structures and Structuralism in Contemporary Philosophy of Mathematics'. *Synthese*, 125: 341–83.

Reisch, George A. (1991). 'Did Kuhn Kill Logical Empiricism?'. *Philosophy of Science*, 58 (2): 264–77.

Reisch, George A. (2005). *How the Cold War Transformed Philosophy of Science: To the Icy Slopes of Logic*. Cambridge: Cambridge University Press.

Resnik, Michael (1981). 'Mathematics as a Science of Patterns: Ontology and Reference'. *Nous*, 15: 529–50.

Resnik, Michael (1982). 'Mathematics as a Science of Patterns: Epistemology'. *Nous*, 16: 95–105.

Resnik, Michael (1990). 'Between Mathematics and Physics'. *PSA: Proceedings of the Biennial Meeting of the Philosophy of Science Association*, 1990 (2): 369–78.

Resnik, Michael (1997). *Mathematics as a Science of Patterns*. Oxford. Clarendon Press.

Reynhout, Kenneth (2011). 'Alain Badiou: Hidden Theologian of the Void?' *The Heythrop Journal*, 52(2): 219–33.

Rickles, Dean and French, Steven and Saatsi, Juha (eds) (2006). *The Structural Foundations of Quantum Gravity*. Oxford: Oxford University Press.

Riera, Gabriel (ed.) (2005). *Alain Badiou: Philosophy and its Conditions*. Albany, NY: SUNY Press.

Robson, Eleanor and Stedall, Jacqueline (eds) (2009). *The Oxford Handbook of History of Mathematics*. Oxford: Oxford University Press.

Roffe, Jon (2006). 'Book Review: Alain Badiou's *Being and Event*'. *Cosmos and History*, 2(1–2): 327–38.

Roffe, Jon (2012). *Badiou's Deleuze*. Montreal: McGill-Queen's University Press.

Rosenberg, Alex and Balashov, Yuri (eds) (2002). *Philosophy of Science: Contemporary Readings*. London and New York: Routledge.

Ross, Don (2008). 'Ontic Structural Realism and Economics'. *Philosophy of Science*, 75 (5): 732–43.

Russell, Bertrand (1931). *The Scientific Outlook*. London: Unwin Brothers Ltd.

Russell, Bertrand (1992). *The Analysis of Matter*. London: Routledge.

Russell, Bertrand (1996). *Our Knowledge of the External World*. London and New York: Routledge.

Russell, Bertrand (2009). *Human Knowledge: Its Scope and Limits*. London: Routledge.

Rutherford, Donald (ed.) (2006). *The Cambridge Companion to Early Modern Philosophy*. Cambridge: Cambridge University Press.
Sacilotto, Daniel (2013) 'Towards a Materialist Rationalism: Plato, Hegel, Badiou'. *The International Journal of Badiou Studies*, 2(1): 36–77.
Salmon, Wesley (1984). *Scientific Explanation and the Causal Structure of the World*, Princeton: Princeton University Press.
Sankey, Howard (2008). *Scientific Realism and the Rationality of Science*. Aldershot: Ashgate.
Sarkar, Sahotra (ed.) (1996). *Logical Empiricism at Its Peak: Schlick, Carnap, and Neurath*. London and New York: Garland Publishing.
Sarukkai, Sundar (2005). 'Revisiting the "Unreasonable Effectiveness" of Mathematics'. *Current Science*, 88(3): 415–23.
Saunders, Simon (2006). 'Are Quantum Particles Objects?' *Analysis*, 66(1): 52–63.
Schaffer, Jonathan (2003). 'Is There a Fundamental Level? *Noûs*, 37(3): 498–517.
Schantz, Richard (ed.) (2002). *What Is Truth?* Berlin and New York: de Gruyter.
Schlick, Moritz (1936). 'Meaning and Verification'. *The Philosophical Review*, 45(4): 339–69.
Sedley, David (1982). 'Two Conceptions of Vacuum'. *Phronesis*, 27(2): 175–93.
Seibt, Johanna (2000). 'Pure Processes and Projective Metaphysics'. *Philosophical Topics*, 101(2–3): 253–89.
Sellars, Wilfrid (1963). *Science, Perception and Reality*. Atascadero, CA: Ridgeview Publishing Company.
Sellars, Wilfrid (1965). 'Scientific Realism or Irenic Instrumentalism'. *Boston Studies in the Philosophy of Science*, 171–204.
Sellars, Wilfrid (1968). *Science and Metaphysics: Variations on Kantian Themes*. London: Routledge & Kegan Paul.
Sellars, Wilfrid (1974). *Essays in Philosophy and Its History*. Dordrecht: Reidel Publishing Company.
Sellars, Wilfrid (1976). 'Is Scientific Realism Tenable? *PSA: Proceedings of the Biennial Meeting of the Philosophy of Science Association*, 1976: 307–34.
Sellars, Wilfrid (1980). *Naturalism and Ontology*. Atascadero, CA: Ridgeview Publishing Company.
Sellars, Wilfrid (1981). 'The Lever of Archimedes'. *The Monist*, 64(1): 3–90.
Shackel, Nicholas (2011). 'The World as a Graph: Defending Metaphysical Graphical Structuralism'. *Analysis*, 71(1): 10–21.
Shapin, Steven (1996). *The Scientific Revolution*. Chicago: The University of Chicago Press.
Shapiro, Stewart (1983). 'Mathematics and Reality'. *Philosophy of Science*, 50(4): 523–48.
Shapiro, Stewart (1997). *Philosophy of Mathematics: Structure and Ontology*. Oxford: Oxford University Press.
Shapiro, Stewart (2000). *Thinking about Mathematics: the Philosophy of Mathematics*. Oxford: Oxford University Press.
Shapiro, Stewart (ed.) (2005). *The Oxford Handbook of Philosophy of Mathematics and Logic*. Oxford: Oxford University Press.
Simmons, J. Aaron and Aikin, Scott F. (2012). 'Prospects for a Levinasian Epistemic Infinitism'. *International Journal of Philosophical Studies*, 20(3): 437–60.
Smart, J.J.C. (1963). *Philosophy and Scientific Realism*. London: Routledge & Kegan Paul.

Smith, Nicholas H. (ed.) (2002). *Reading McDowell: On Mind and World*. London and New York: Routledge.

Snyder, Laura J. (2006). *Reforming Philosophy: A Victorian Debate on Science and Society*. Chicago: The University of Chicago Press.

Solomonoff, Ray J. (1964a). 'A Formal Theory of Inductive Inference. Part I'. *Information and Control*, 7: 1–22.

Solomonoff, Ray J. (1964b). 'A Formal Theory of Inductive Inference. Part II'. *Information and Control*, 7: 224–54.

Spinoza, Baruch (2002). *Complete Works*. Indianapolis, IN and Cambridge, MA: Hackett.

Stapleton, Timothy J. (ed.) (1994). *The Question of Hermeneutics*. Dordrecht: Kluwer Academic Publishers.

Steiner, Mark (1973). 'Platonism and the Causal Theory of Knowledge'. *The Journal of Philosophy*, 70(3): 57–66.

Steiner, Mark (1989). 'The Application of Mathematics to Natural Science'. *The Journal of Philosophy*, 86(9): 449–80.

Steiner, Mark (1998). *The Applicability of Mathematics as a Philosophical Problem*. Cambridge, MA and London: Harvard University Press.

Strawson, P.F. (2003) [1959]. *Individuals: An Essay in Descriptive Metaphysics*. London and New York: Routledge.

Tahko, Tuomas E. (2012). *Contemporary Aristotelian Metaphysics*. Cambridge: Cambridge University Press.

Tegmark, Max (2007). 'The Mathematical Universe'. *Foundations of Physics*, 38(2): 101–50.

Teller, Paul (1995). *An Interpretive Introduction to Quantum Field Theory*. Princeton, NJ: Princeton University Press.

Tho, Tzuchien and Bianco, Giuseppe (eds) (2013). *Badiou and the Philosophers*. London: Bloomsbury.

Tiles, Mary (1984). *Bachelard: Science and Objectivity*. Cambridge: Cambridge University Press.

Toscano, Alberto (2000). 'To Have Done with the End of Philosophy'. *Pli: The Warwick Journal of Philosophy*, 9: 220–38.

Toulmin, Stephen (1961). *Foresight and Understanding: An Enquiry into the Aims of Science*. New York and Evanston, IL: Harper Torchbooks.

Turing, Alan (1936). 'On Computable Numbers, with an Application to the Entscheidungsproblem'. *Proceedings of the London Mathematical Society*, 2nd series, 42, 230–65.

Turing, Alan (1950). 'Computing Machinery and Intelligence'. *Mind*. 59(236): 433–60.

Uebel, Thomas E. (1992). *Overcoming Logical Positivism from Within: The Emergence of Neurath's Naturalism in the Vienna Circle's Protocol Sentence Debate*. Amsterdam: Rodopi.

van Fraassen, Bas C. (1975). 'Wilfrid Sellars on Scientific Realism'. *Dialogue*, 14(04) (May 5): 606–16.

van Fraassen, Bas C. (1976). 'On the Radical Incompleteness of the Manifest Image'. *PSA: Proceedings of the Biennial Meeting of the Philosophy of Science Association*, 1976: 335–43.

van Fraassen, Bas C. (1980). *The Scientific Image*. Oxford: Clarendon Press.

van Fraassen, Bas C. (1991). *Quantum Mechanics: An Empiricist View*. Oxford: Oxford University Press.

van Fraassen, Bas C. (2002). *The Empirical Stance*. New Haven, CT: Yale University Press.

van Fraassen, Bas C. (2006). 'Structure: Its Shadow and Substance'. *The British Journal for the Philosophy of Science*, 57(2): 275–307.

van Fraassen, Bas C. (2008). *Scientific Representation*. Oxford: Oxford University Press.

Vardoulakis, Dimitri (ed.) (2011). *Spinoza Now*. Minneapolis, MN: University of Minnesota Press.

Varzi, Achille (2005). *Ontologia*. Roma and Bari: Laterza.

Votsis, Ioannis (2004). *The Epistemological Status of Scientific Theories: An Investigation of the Structural Realist Account*. PhD thesis, London School of Economics and Political Science.

Votsis, Ioannis (2005). 'The Upward Path to Structural Realism'. *Philosophy of Science*, 72: 1361–72.

Wang, Hao (1996). *A Logical Journey: From Gödel to Philosophy*. Cambridge, MA: MIT Press.

Warner, Rob (2010). *Secularization and Its Discontents*. London: Continuum.

Watkin, Christopher (2011). *Difficult Atheism: Post-theological Thinking in Alain Badiou, Jean-Luc Nancy and Quentin Meillassoux*. Edinburgh: Edinburgh University Press.

Watkins, Eric (ed.) (2013). *The Divine Order, The Human Order, and the Order of Nature*. Oxford: Oxford University Press.

Weyl, Hermann (1949). *Philosophy of Mathematics and Natural Science*. Princeton, NJ. Princeton University Press.

Whewell, William (1862). *Astronomy and General Physics, considered with reference to Natural Theology*. London: Bradbury and Evans.

Whitehead, Alfred North (1948) [1925]. *Science and the Modern World*. New York: Pelican Mentor.

Wigner, Eugene (1960). 'The Unreasonable Effectiveness of Mathematics in the Natural Sciences'. *Communications in Pure and Applied Mathematics*, 13(I): 1–14.

Wilholt, Torsten (2006). 'Lost on the Way from Frege to Carnap: How the Philosophy of Science Forgot the Applicability Problem'. *Grazer Philosophische Studien*, 73: 69–82.

Williamson, Timothy (2011). 'What is Naturalism?' *The New York Times Opinionator*. Online, http://opinionator.blogs.nytimes.com/2011/09/04/what-is-naturalism/ (accessed 1 May 2014).

Wilson, Catherine (1995). *The Invisible World: Early Modern Philosophy and the Invention of the Microscope*. Princeton, NJ: Princeton University Press.

Wilson, Mark (2000). 'The Unreasonable Uncooperativeness of Mathematics in the Natural Sciences'. *The Monist*, 83(2): 296–314.

Wilson, Mark (2006). *Wandering Significance: An Essay on Conceptual Behavior*. Oxford: Clarendon Press.

Wimsatt, William (2007). *Re-engineering Philosophy for Limited Beings: Piecewise Approximations to Reality*. Cambridge, MA: Harvard University Press.

Wittgenstein, Ludwig (1976). *Lectures on the Foundations of Mathematics, Cambridge 1939*. Edited by Cora Diamond. Cambridge: Cambridge University Press.

Wolfram, Stephen (2002). *A New Kind of Science*. Champaign, IL: Wolfram Media.

Worrall, John (1989). 'Structural Realism: The Best of Both Worlds?' *Dialectica*, 43(1–2): 99–124.

Worrall, John (1994). 'How to Remain (Reasonably) Optimistic: Scientific Realism and the "Luminiferous Ether"'. *PSA: Proceedings of the Biennial Meeting of the Philosophy of Science Association*, 1: 334–42.

Wright, Crispin (1992). *Truth and Objectivity*. Cambridge, MA: Harvard University Press.

Zahar, Elie (2001). *Poincaré's Philosophy: From Conventionalism to Phenomenology*. Chicago and La Salle, IL: Open Court.

Zalamea, Fernando (2012). *Synthetic Philosophy of Contemporary Mathematics*. Translated by Z. L. Fraser. Falmouth and New York: Urbanomic and Sequence Press.

Zenil, Hector (ed.) (2012). *A Computable Universe: Understanding and Exploring Nature as Computation*. Singapore: World Scientific.

Zhang, Jiajie, and Donald A. Norman (1994). 'Representations in Distributed Cognitive Tasks'. *Cognitive Science*, 122: 87–122.

Zhmud, Leonid (2012). *Pythagoras and the Early Pythagoreans*. Oxford: Oxford University Press.

Žižek, Slavoj (2006). *The Parallax View*. Cambridge, MA and London: MIT Press.

Index